DATE DUE

APR 2 4 1995	
MAR 1 1996	
MAR 3 1 1997	
MAR 1 9 1998	

The Green Rainbow

The Green Rainbow

ENVIRONMENTAL GROUPS

IN WESTERN EUROPE

Russell J. Dalton

Yale University Press   New Haven & London

Published with the assistance from the Louis Stern Memorial Fund.

Designed by Deborah Dutton.
Set in Times Roman text and Futura Condensed display types by DEKR Corp., Woburn, Massachusetts.
Printed in the United States of America by Vail-Ballou Press, Binghamton, New York.

Library of Congress Cataloging-in-Publication Data

Dalton, Russell J.
    The Green rainbow : environmental groups in Western Europe /
Russell J. Dalton.
        p.   cm.
    Includes bibliographical references and index.
    ISBN 0–300–05962–0 (alk. paper)
    1. Environmentalism—Europe.   2. Pressure groups—Europe.
3. Social ecology—Europe.   I. Title.
GE199.E85D35   1994
363.7'05763'094—dc20   94-4956
                                                                                            CIP

A catalogue record for this book is available from the British Library.

The paper in this book meets the guidelines for permanence and durability of the Committee on Production Guidelines for Book Longevity of the Council on Library Resources.

10 9 8 7 6 5 4 3 2 1

TO DORIS P.

Contents

Part 4 Political Repertoires

Part 5 Conclusion

CONTENTS

Tables and Figures

Tables

Figures

Preface

At the beginning of this project, many argued that green politics was a passing novelty in contemporary politics. Today it seems everyone is green, or at least claims to be. What has produced this greening of European politics, and what is the future of the green movement?

This is, first, a study to examine the development of the green movement in Western European democracies. In it I attempt to describe environmental interest groups as important new participants in the contemporary political process. How do these groups conceptualize their political role and carry out their activities? What resources do they command, and what goals do they pursue?

In addition, this research marks an evolution of my interest in the political changes transforming advanced industrial societies. After conducting considerable research on the changing values of Western publics, which documents the growing interest in postmaterial issues, I looked for ways to validate this public opinion evidence with concrete political examples. If advanced industrial societies were really changing, where was the evidence in the political process or in the operation of the political system? I focused on the environmental movement for several reasons. The green movement is generally considered the archetypical example of postmaterial politics; a study of the movement could determine whether these claims are justified. Furthermore, environmental interest groups should play a key role in translating the public's changing policy interests into a political force. As the green movement has developed, these groups have helped to clarify the goals and identity of the movement—to define what it means to be green. More so than green parties or environmental policy studies, research on environmental groups can explore the link between the postmaterialist citizenry and the political process. Finally, a growing academic interest in social movements by sociologists and political scientists enabled me to locate this research in a rich and rapidly expanding area of scholarship and, I hope, to make a contribution to our understanding of social movements more generally.

A grant from the National Science Foundation (SES 85–10989) provided the support for a research team to collect data from the major environmental groups in Western Europe and to conduct personal interviews with representatives of most of these organizations. I am greatly indebted to Christos Bourdouvalis, Edward DeClair, and Robert Rohrschneider for their assistance in designing this project and in conducting these interviews.

Several other individuals assisted this project along the way. Philip Lowe offered sage advice from his survey of British environmental groups and graciously shared his results with me. Frank Wilson also shared his questionnaire on French interest groups. I hope both will see how their advice positively benefited the research presented here.

The assistance of the European Environmental Bureau (EEB) and its director, Ernst Klatte, was an invaluable aid in contacting environmental interest groups and gaining their cooperation in this study. The EEB also allowed us to attend its annual meeting in Brussels, which provided extensive opportunities to meet with group officials and to observe their political discussions firsthand.

My own thinking on social movements greatly benefited from the collaboration with Wilhelm Bürklin and Manfred Kuechler in hosting an NSF-funded international conference on new social movements; the conference grant also partially supported this research (INT 85–21364). This conference produced a comparative study of social movements in advanced industrial societies (Dalton and Kuechler, 1990). Both Willi and Manfred have been generous in their advice, and the contributors to the conference enriched my understanding of social movement theory and practice. Many of the specific research hypotheses developed in this book had their genesis at this meeting.

I also wish to thank those colleagues who offered criticism on the manuscript or chapters of the book: Christos Bourdouvalis, Mario Diani, J. Craig Jenkins, Sheldon Kamienecki, Herbert Kitschelt, Lester Milbrath, Thomas Rochon, Robert Rohrschneider, Wolfgang Rüdig, and the anonymous reviewers for Yale University Press. I am also indebted to several of my colleagues at the University of California, Irvine, for convincing me to place more emphasis on the "t-word." The book is much improved because of their comments. John Covell of Yale University Press has been patient and supportive in seeing this project through to publication. I wish to thank Karen Gangel for the exceptional time and care she invested in editing the manuscript. Cheryl Larsson expertly created the artwork for the book, and Alexa Cole created the index. Ginny, Penn, Kenny, and Biff were also very understanding of the time it took to complete this work.

Finally, I should like to give special thanks to the representatives of the environmental interest groups who consented to be interviewed and participate in this project. Although some colleagues warned of possible difficulties in gaining access to these organizations, we were almost uniformly received with open arms and a willingness to discuss sensitive (and sometimes compromising) issues. This interchange was a truly memorable life experience, as well as an exciting research project. We spent weeks talking with some of the most fascinating and politically motivated people in the green movement; the participants provided a tutorial in green politics. In addition, many of these individuals showed us great kindness and were supportive of the research enterprise: one respondent invited us to interview her at home while she awaited the birth of her child; others assisted by supplying facts and figures for our data collection. Their opinions and evidence are the basis for this study. We hope that in providing a better understanding of green politics and its implications for contemporary democracies we have repaid their efforts.

Abbreviations of Key Terms

Ad1T   Les Amis de la Terre
        Friends of the Earth

BBU   Bundesverband Bürgerinitiativen Umweltschutz
        Federal Association of Citizen Action Groups for Environmental
        Protection

BUND   Bund für Umwelt- und Naturschutz Deutschland
        Germany Federation for Environmental and Nature Protection

CoEnCo Council for Environmental Conservation

COLINE Comité Législatif d'Information Ecologique
        Legislative Committee for Ecological Information

CPRE   Council for the Protection of Rural England

DBV   Deutscher Bund für Vogelschutz
        German Federation for Bird Protection

DN   Danmarks Naturfredningsforening
        Danish Association for Nature Conservation

DNR   Deutscher Naturschutzring
        German Nature Protection Ring

DOF   Dansk Ornitologisk Forening
        Danish Ornithological Society

EC   European Communities

EEB   European Environmental Bureau

FFPS   Fauna and Flora Preservation Society

FFSPN	Fédération Française des Sociétés de Protection de la Nature
	French Federation of Societies for the Protection of Nature
FoE	Friends of the Earth
HSPN	Hellenic Society for the Protection of Nature
IPEE	Institut pour une Politique Européene de l'Environnement
	Institute for European Environmental Policy
ISA	Ideologically Structured Action
IUCN	International Union for the Conservation of Nature
IVN	Institut voor Natuurbeschermingseducatie
	Institute for Nature Conservation Education
LIPU	Lega Protezione di Uccelli
	League for the Protection of Birds
LMO	Landelijk Milieu Overleg
	National Platform for the Environment
NOAH	An alternative environmental group in Denmark
NSM	New Social Movement
RM	Resource mobilization
RSNC	Royal Society for Nature Conservation
RSPB	Royal Society for the Protection of Birds
SERA	Socialist Environmental Resources Association
SME	Stichting Milieu-edukatie
	Foundation for Environmental Education
SNM	Stichting Natuur en Milieu
	Foundation for Nature and the Environment
SNPN	Société Nationale de Protection de la Nature
	National Society for the Protection of Nature
TCPA	Town and Country Planning Association
UNEP	United Nations Environmental Program
VMD	Vereniging Milieudefensie
	Society for the Defense of the Environment
WWF	World Wildlife Fund

1

INTRODUCTION

Chapter One

ENVIRONMENTALISM AND

SOCIAL MOVEMENT THEORY

Greenery is now sprouting all over the political landscape. Environmental activists criticize the excesses of today's consumer-driven society and demand a restructuring of life-styles and the economic system. Conservationists lament the rapid destruction of the ecosphere and the accelerating rates of species loss and environmental damage. The protests of citizens' groups draw public attention to the emerging environmental problems of advanced industrial societies: thousands demonstrate against acid rain and the consequent destruction of Europe's forests; the little rubber boats of Greenpeace dauntlessly battle the dumping of waste at sea; wildlife organizations campaign with equal vehemence for the protection of African elephants and the English hedgehog; and local protest groups spring up like mushrooms around nuclear power installations. The representatives of newly formed green parties castigate all of the established parties for their duplicity in this environmental crisis and bring these views into the legislative process. These organizations, their members, and the issues they address represent the core of the environmental movement.

The environmental movement has rapidly become an important, and contentious, element in the political system of most Western democracies. Because the movement challenges many of the norms and institutions of the dominant political order, it is a source of political controversy and polarization. To its supporters, the green movement is seen as the vanguard of a new society; to its critics, environmentalism represents a fundamental threat to the established socioeconomic system. Indeed, the rise of the green movement has sparked much political controversy, electoral conflict, and resistance by the established political elite.

At one level, we are interested in environmentalism because of its newly achieved role in contemporary politics. The size of the movement and the electoral impact of green parties alone are sufficient to warrant the attention of scholars and political analysts. Nearly all Western democracies now have ministries devoted primarily to environmental issues. With increasing frequency, government leaders and opposition figures espouse support for environmental issues to court the favor of green voters, although their environmental advocacy often reflects more symbolic gesturing than substance. Illustrating this broadening political concern for green issues, the Commission of the European Communities approved its Fourth Environmental Action Plan in 1986. The commission designated 1986–87 as the European Year of the Environment, aimed at "convincing every element in our societies . . . that protection of the environment must become a fundamental element in their lives and all human activities." Two years later, the commission established a European Environmental Agency to coordinate the environmental programs of the member states. In 1992, European environmentalists and government representatives were the driving forces behind the global environmental summit at Rio de Janeiro. The European environmental movement is coming of age.

At a second level, the interest in this topic reflects a belief that environmentalism typifies the new political controversies and political style that are transforming the politics of advanced industrial societies. Analysts frequently describe environmentalism as an example of the changing value priorities of Western publics. As a growing proportion of the public begins to emphasize postmaterial values, the composition of the political agenda shifts from traditional economic and security concerns to the noneconomic and quality-of-life values of a postindustrial society (Inglehart 1977, 1990; Cotgrove 1982; Milbrath 1984). Opinion surveys in Western Europe and North America document widespread public concern with environmental quality and public endorsement for policies of environmental reform. When given the choice between further economic growth and protecting the environment, most citizens now choose the environment![1] After decades of discussion, Western publics are displaying an awareness that present life-styles and economic patterns cannot continue unchecked and that environmental reform is necessary to ensure the present quality of life.

Social scientists also cite the environmental movement as an example of a new style of citizen politics, composed of public interest lobbies and elite-challenging forms of political action (Lowe and Goyder 1983; Dalton 1988; Berry 1977). By many accounts, the members of the various European environmental groups now outnumber the formal membership in political parties. And when the established parties have proven unresponsive to calls for environmental legislation, new green parties have formed to carry these themes into the electoral arena. By the beginning of the 1990s, green parties or their New Left supporters had won seats in the national parliament or EC parliament for most nations in Western Europe.

The goals and style of the environmental movement are not a singular challenge to the political order of advanced industrial societies; many of the same characteristics are found in other new social movements, such as the women's movement, peace groups, and the self-help movement (Dalton and Kuechler 1990; Klandermans et al. 1988; Rucht 1992). In a few short years, the women's movement has stimulated a fundamental change in basic social roles that had existed for centuries (Freeman 1975; Gelb 1989; Katzenstein and Mueller 1987). Similarly, contemporary peace groups have often mobilized widespread popular support and indirectly influenced government policy options on defense issues (Rochon 1988; Klandermans, ed., 1991). In many instances, the political style, organizational tendencies, and even personnel of these various movements display considerable overlap. Collectively, these movements are calling for a new agenda for contemporary societies and demanding that governments open the political process to more diverse and citizen-oriented interests.

This is, first of all, a study of the growth and development of the environmental movement in European political systems. At the same time, however, we believe that research on the environmental movement has the promise of connecting abstract theories about the nature of postindustrial politics—postmaterial values, theories of organizational transformation, and changing political styles—to the changes actually occurring in these societies. Building on prior social movement theory, we argue that the ideology of environmentalism is what produces these new patterns of political action. Moreover, by reintroducing ideology to the study of social movements, we are better able to explain their goals and methods; and we maintain that ideology is the connecting mechanism that produces what is new about environmentalism and other new social movements. The result of these efforts should be a better understanding of behavior associated with such movements and the nature of citizen politics in advanced industrial societies.

Resource Mobilization and New Social Movements

Over the past two decades, most research on social movements has been driven by one of two contrasting models of social movement behavior: resource mobilization theory (RM) and what has been termed New Social

Movement theory (NSM).[2] Both theoretical models represent attempts to address the characteristics of the new citizen action groups and public interest lobbies that prior models had not adequately captured (Dalton and Kuechler 1990, chap. 1). These two theories provide the starting point for our exploration of the European environmental movement.

Resource mobilization theory was developed by a group of American sociologists (Oberschall 1973; McCarthy and Zald 1977; Gamson 1975; Tilly 1978; Jenkins 1983; Zald and McCarthy 1987). Theorists of RM presume that the formation of a social movement and its behavior is primarily dependent on the existence of organizations that mobilize resources in pursuit of a cause and determine the activities of the movement. The study of social movement is thus equated with a study of organizational behavior. Among political scientists, many of the theoretical interests and predictions of the resource mobilization paradigm are paralleled in the rational-choice model of political groups (Olson 1965; Frohlich et al. 1971; Moe 1980; Chong 1991).[3] Resource mobilization/rational-choice theory shifts the focus of our attention away from the popular base of social movements and toward the organizational features of the movement. To an extent, this is an economic model of political groups, where organizational needs and opportunities are more important than individual sentiments and political values in generating a political group and guiding its actions.

RM/rational-choice theory provides a powerful framework for explaining and predicting the behavior of social movement organizations (SMOs). Resource mobilization theorists presume, for example, that political dissatisfaction and social conflict are inherent in every society. Therefore, the creation of a social movement organization is not primarily a reflection on the existence of political grievances in society; instead, it depends on the presence of sufficient resources and entrepreneurial expertise to create and sustain the movement. The recent proliferation of public interest groups and citizens lobbies is a by-product of the increasing number of potential entrepreneurs who possess the organizational skills necessary to create a movement organization and the greater availability of resources to sustain these organizations (McCarthy and Zald 1977, 1224; Downing and Brady 1974; Frohlich et al. 1970; Moe 1980).[4]

Once a social movement organization (SMO) is formed, RM/rational-choice theory holds that the behavior of the organization is shaped by pragmatic calculations of how to meet its resource needs and make maximal use of the available opportunity structures (Zald and McCarthy 1987; McCarthy and Zald 1973). This organizational planning means that an SMO bases its tactics on conscious calculations of how best to advance the goals of the organization—whether through dramatic protests or quiet political lobbying. As part of this calculus of group action, the original political goals of a movement must compete with the organization's desire (represented by its leadership and staff) to maintain itself. Some of the activities of an SMO are therefore directed at organizational maintenance (servicing the membership and mobilizing new support) and not just policy influence. A citizen group, for example, might plan a demonstration or stage a dramatic event not to influence policy makers but to generate increased public support (and financial contributions) for the organization.

The RM approach also leads us to examine social movement organizations as embedded in a network of formal and informal political relationships (e.g., Morris 1984; Klandermans 1990; Diani 1994). The strategic and resource needs of SMOs stimulate them to establish alliances with other political actors (established interest groups, political parties, and the like) to ensure a sufficient level of resources to survive and to facilitate access to the political process. Paul Downing and David Brady (1974) have illustrated that the support of private foundations and public grants was vital during the formative stages of many American environmental groups (see also Walker 1991, chap. 5); Edward Walsh's (1989) research outlines the broad political ties antinuclear power groups developed in the wake of Three Mile Island; Hanspeter Kriesi and Philip van Praag (1987) similarly explore alliance patterns between old and new movement networks in Holland. The organization that wants to be effective will seek support from outside the movement, and RM emphasizes this pragmatic pattern of alliance building.

The resource needs of an SMO further influence the internal structure of the organization (see chap. 4). In contrast to the possibly amorphous nature of the underlying social movement, the theory maintains that SMOs are inclined to adopt a hierarchical and highly routinized structure to maximize their efficiency in collecting money, activating members, and achieving policy success (Gamson 1975; Tilly 1978). Stephan Barkan's (1979) research apparently validated this logic by documenting the fundamental organizational problems encountered by American antinuclear power groups that did not adopt this structure. In many small citizen groups, this concentration of authority frequently occurs in the hands of an entrepreneur who creates and directs the organization.[5] Many SMOs would not exist without the initiative of a single individual or small group of people, even though public interest in the cause may be long-standing.

The RM/rational-choice approach thus maintains that organizational and entrepreneurial resources combine with political opportunities (created by elite competition, electoral politics, and the patterns of political alliances) to generate mass movements. The resource needs and opportunities of a SMO affect its structure, tactics, patterns of political alliance, and goals (Zald and McCarthy 1987; McCarthy and Zald 1977; McAdam 1988). This model provides an integrated theory of how organizations are formed, how public support is mobilized, how organizational behavior is developed, and how political tactics are decided.

Although RM/rational-choice theory has advanced social movement research, there has been a growing awareness of the limits of this approach (Melucci 1984; Klandermans 1986; Klandermans et al. 1988; Morris and Mueller 1992). One impetus for this reevaluation was a concern that in enshrining rational action, the RM approach removed politics and values from the study of mass movements. Past applications of the theory were frequently indifferent to the ideological content of a movement; the theory was applied in an almost mechanist way to a diversity of organization—labor unions, civil rights groups, farm workers, environmental groups, right-wing extremists, antibusing groups, the sanctuary movement, women's groups, Mothers Against Drunk Driving (MADD)—without incorporating the political nature of each group into the workings of the model. This generic approach dilutes the definition of a social movement to mean almost any mass-membership association—the term *social movement* implies a popular movement directed at significant social or political reform. Furthermore, it misses the significant variation in behavior that occurs between social movements and social movement organizations because of their differing goals and values.

A second impetus for this reevaluation was the rise of a new set of contemporary citizen movements—environmental groups, women's groups, the peace lobby, and self-help groups—whose nature and behavior appear inconsistent with RM theory (Brand 1982; Raschke 1985; Brand et al. 1986; Dalton and Kuechler 1990; Rucht 1992). Although economic and class interests frequently served as the basis of earlier social movements (especially those studied within the rational-choice framework), these new movements are primarily concerned with collective goods that involve social, cultural, and quality-of-life issues. Furthermore, some new environmental organizations—such as Friends of the Earth or the Conservation Society in Britain—seem to represent a new style of internal politics and political action that differs from the pattern of earlier social movements or voluntary associations, including older environmental groups.[6] German sociologists coined the term *Neue soziale Bewegungen* to describe this phenomenon (Brand 1982; Brand et al. 1986), and the term *New Social Movement* entered the English research vocabulary as an identifier for this new type of citizen movement.

The conceptual framework for studying NSM is often as ill-defined and imprecise as the groups being analyzed, but this literature describes several characteristics observed within these movements that differ in fundamental ways from the expectations generated by RM/rational-choice theory. For example, rather than a centralized and hierarchically structured organization as implied by RM/rational-choice theory, sectors of the environmental movement seemingly prefer a decentralized structure that reflects the participatory tendencies of their members. While the resource mobilization literature treats such an organizational style as dysfunctional and harmful to the effectiveness of the movement, the NSM literature maintains that this style actually serves as an attractive force for potential movement supporters (Downey 1986; Kitschelt and Hellemans 1990).

Instead of accepting the central role of leadership and the political entrepreneur, the ethos within NSMs supposedly evokes an aversion to the elitism and hierarchy that RM/rational-choice theory maintains is a prerequisite for an effective SMO. Specialists cite the Green Party of Germany (Die Grünen) as the prime illustration of these antiorganizational tendencies (e.g., Poguntke 1988; Kitschelt 1989; Kitschelt and Hellemans 1990; Frankland and Schoonmaker 1993). Although the actions of the party often fall short of its rhetoric, especially recently, many party militants display an almost pathological aversion to oligarchy and majoritarian decision making, with some Green activists explicitly citing Roberto Michels's "iron law of oligarchy" as the sin they strive to avoid (Kitschelt 1989). The European peace movement displays similar antioligarchic tendencies (Mushaben 1989; Rochon 1988), as does the women's movement (Ferree 1987; Gelb 1989).

The NSM literature further depicts a distinctly different picture of the political tactics of these SMOs. Several researchers claim that NSM organizations consciously remain outside the network of established political alliances that resource mobilization theory stresses as a path toward organizational effectiveness. NSMs supposedly reject close ties to the dominant political groups because they fear that cooperation will lead to co-optation and ineffectiveness. Indeed, we shall describe numerous instances of these suspicions becoming reality. NSM scholars claim that this pattern of isolation carries over to the political tactics of these movements. Much current literature on European interest groups emphasizes the neocorporatist tendencies of most interest organizations (e.g., Lehmbruch and Schmitter 1982; Grant 1985), and RM/rational-choice theory implies that NSMs will work within this established structure if it provides an effective means of influence. In contrast, the NSM literature suggests that many environmental organizations consciously reject neocorporatism as incompatible with their goals and political style. These antistatist orientations presumably lead New Social Movements to adopt unconventional methods of political action (protests, demonstrations, and spectacular actions) as a standard part of their political repertoire.

Across a range of social movement characteristics, these two approaches yield contrasting expectations about how groups form their internal structure, make their decisions, adopt political tactics, and mobilize individuals to participate. Rather than substantiating either model, research on New Social Movements often yields evidence that partially supports both perspectives.[7] Both apparently tap part of the reality of the new movements of advanced industrial societies, and both provide insights that are useful in understanding the goals and behaviors of these movements. Yet because neither approach alone provides a sufficient explanation, a synthesis of these contrasting models is required.

Ideologically Structured Action

This study of the European environmental movement attempts to reconcile the divergent perspectives of RM theory and the NSM approach to provide an integrated framework for understanding the behavior of environmental groups (and the other New Social Movements of advanced industrial societies). The choice of environmental interest groups as a focal point for this study underlines the importance we attribute to these organizations in defining the nature of the green movement. Widespread popular support for environmental reform is not sufficient to ensure that public policy will change; organized and sustained activity is a virtual prerequisite for policy influence in contemporary democracies. Once formed, environmental interest groups confront many of the organizational pressures noted in RM theory. These groups must mobilize sufficient resources to ensure their survival, and these resource needs are related to the strategy and tactics of the organizations. We shall, for instance, discuss how the organizational pressures of RM theory are apparent in the institutional structure, decision-making patterns, and political tactics of environmental interest groups.

What is distinctive about this research is our attempt to show how the ideology or collective identity of an SMO can independently affect its structure and behavior. At a theoretical level, certainly, ideology is central to the study of social movements. Charles Tilly, for example, sees ideology and political beliefs as a defining element of social movements; for him, social movements are "a group of people identified by their attachment to some particular set of beliefs" (Tilly 1978, 9). Similarly, Daniel Foss and Ralph Larkin (1986, 2) define a social movement as a group of people who share interests that are incompatible with the existing social and political order (see also McCarthy and Zald 1977, 1217; Zald and McCarthy 1987, 20; Smelser 1963). In fact, most studies of social movements begin by discussing the underlying political values and beliefs, but researchers do not incorporate this attention to ideology into their analyses.[8]

Our research represents an effort to reintroduce this missing ideological element into the study of social movements. We can view RM theory as a value-free model of organizational behavior that should occur if groups were solely concerned with maximizing their short-term outputs. We would argue that the underlying, though often unarticulated, premise of the New Social Movement literature is that the specific ideological orientation of these movements leads them to deviate from the value-free patterns of behavior represented by RM theory. The distinct political values of the core activists and the history of an organization define its political identity, which then serves as the basis for attracting a certain type of membership, projecting an image of the group to potential allies and opponents, and making the strategic and tactical decisions of the organization.[9]

These considerations lead us to propose a general framework of *Ideologically Structured Action* (ISA) that guides our study of European environmental groups (fig. 1.1). This framework presumes that virtually all political organizations, even New Social Movements, eventually recognize the organizational constraints embodied in resource mobilization theory. In principle, however, several options are normally available to address these organizational needs—and political concerns derived from the ideology and identity of a group determine the specific options available to a group and its choice between the available options.[10]

The ideology of founders and the behavior of the group create an identity, both for its participants and the outside world. This identity involves the broad goals of the organization and its conceptualization of the political world; we will describe these environmental identities in chapter 2.[11] As portrayed in figure 1.1, a group's identity should have immediate consequences for the types of resources that the group can potentially mobilize. Even if all movement organizations need financial support, the nature of a group obviously affects where it is likely to find contributors. The World Wildlife Fund, for example, may successfully collect corporate donations and mount joint promotion campaigns with business enterprises; this option is not open to Robinwood, Earth First! and other radical environmental groups. Similarly, the ideology and identity of a political group should influence its organizational structure. In the case of New Social Movements,

FIGURE 1.1
A Model of Ideologically Structured Action

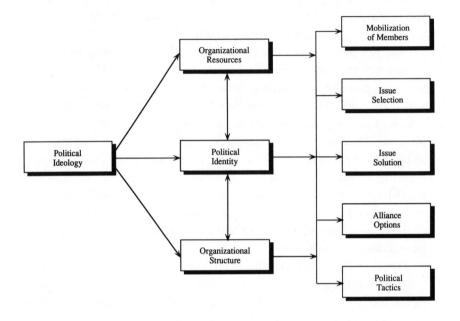

their emphasis on participatory politics and direct democracy should encourage these groups to develop such practices within their own organizations.

The ideology and identity of a political organization are more than just a blueprint for organizing and securing funds, however. The ideology of a group, like that of an individual citizen, provides a framework for organizing and interpreting the political world; it defines core values and peripheral concerns. For an environmental group, these core values might be a concern for life and the aesthetics of nature; for the labor movement it might be maximizing the life chances of its members. An ideology also provides a framework for interrelating discrete phenomena and making sense of the complexities of politics. In policy debates one often discerns that what seems to be sufficient policy reform for one group appears insufficient to another or regressive to a third. In a sense, the meaning of political actions is a constructed reality rather than an objective fact upon which all agree. Bert Klandermans goes so far as to argue that for social movement organizations *"interpretations of reality rather than reality itself guide political actions"* (Klandermans 1991, 8).

Once a social movement organization is established and secures the resources needed to begin operation, its identity influences the factors listed in the right column of figure 1.1: the means of attracting supporters, perceptions of environmental issues, and the political options that are open to it. For instance, a group's environmental ideology and organization mobilize members by creating an identity that appeals to potential supporters. Bert Klandermans (1988) refers to this as consensus mobilization, in which a movement organization presents specific interpretations of political events that reflect the group's environmental orientation and might attract new supporters.[12] Greenpeace, for instance, is selling a distinctive image of environmentalism with its combination of issues and tactics. Greenpeace is telling potential supporters what the group stands for and those who contribute to the organization know what they are endorsing. In less dramatic fashion, the Royal Society for Nature Conservation is also selling its image of environmentalism to its supporters.

A group's ideology also provides the focus for political action. David Snow and Robert Benford (1988) describe this as diagnostic framing. An ideology determines what political goals are most important and what objectives the group should pursue. This is most obvious in the issue agenda of environmental groups: some are concerned with questions of nuclear power, others with bird protection. In addition, the diagnosis can include judgments about the sources of the problem and the required reforms. One environmental group may see nuclear power as a simple question of economics and sufficient safety regulation; to another, the same issue may reflect a more fundamental conflict over the distribution of economic and political power within society. Clearly, these different diagnoses lead to different interpretations of the issue and the prospects for reform.

Finally, the group identity of an SMO, once established, affects the choice of political tactics available to it. Snow and Benford (1988) refer to this as prognostic framing. On one hand, a group's identity partially defines the opportunity structures it faces: the likelihood of alliance with (or opposition from) other social and political interests, governmental reactions to its actions, and response from media. On the other hand, that identity may itself define which political tactics are considered appropriate for the group, independent of which method is most likely to achieve the desired results. Just as it would be out of character for the Civic Trust to mount a spectacular mass protest to increase funding for historic sites in Britain, it is difficult to imagine officials from the antisystem Robinwood engaged in neocorporatist negotiations with German business and labor interests.

In short, political opportunity structures are not a fixed characteristic of a political system, as many RM theorists imply, but vary across organizations as a function of their ideological position. Thus, there are probably greater similarities in the organization and functioning of two radical environmental groups in different nations than there are between a radical environmental group and nature conservation group within the same nation. As Jack Walker observes: "Once an interest group is established and manages to secure the resources it needs to operate, it finds that it has narrowed the number of pathways to influence that are open to it" (Walker 1991, 16).

The importance of extending analyses of SMO behavior to include ideology can be illustrated with a brief example. A framework of Ideologically Structured Action maintains that although the Royal Society for the Protection of Birds (RSPB) and Greenpeace are both concerned with issues involving wildlife protection, their disparate environmental ideologies lead to large and predictable differences in political opportunities and behavior. A group such as the RSPB, which accepts the dominant social order (and is actually drawn from the elite of society), is more likely to receive support and build alliances with business interests, private foundations, and members of the social and political establishment; a challenging group such as Greenpeace may have to find its allies among other movement groups or dissident factions of the establishment. Similarly, the RSPB should have ready access to the political process, whereas Greenpeace must place greater reliance on unconventional methods of political influence. In a like manner, the identity of these two groups should lead them to attract different memberships reflecting distinct values and methods.

Although the creation and projection of an ideology for a group or a movement is an important research concern, we do not attempt to specify all of the existing ideological currents and subcurrents in the green movement (see, e.g., Bramwell 1989; Eckersly 1992). Instead, our goal is to use the broad environmental orientations that are recognized within the environmental movement to explain the actions of environmental interest groups.

In spite of the importance of ideology, a purely ideological or psychological model of social movements would mark a regression of theory to the early collective behavior and relative deprivation approaches, because the structural elements of movement organizations would be lacking. Integrating ideology into the structural analysis of RM theory, however, can yield a fuller and more accurate model of the behavior of political organizations. Furthermore, the impact of a given ideology obviously will vary according to its nature. The ideology of some social movements—such as religious fundamentalism or right-wing extremism—might encourage the centralization of authority and an identity based on intolerance. Similarly, the ideology of most nationalist movements will lead them to reject existing political institutions and, thus, traditional methods of political lobbying. Self-help movements will lean toward participatory political structures. In short, the interaction between ideology and the forces of RM theory is vital to explaining the goals and actions of SMOs.

One could rightly note that this concept of ISA simply represents the addition of ideology to the prevailing resource-based model of social movements. Indeed, it seems obvious that ideological considerations should be central to the study of movements that are directed toward significant social or political reform. Ideology provides the reason and logic for social movements, and movements can be understood only in reference to this framework. Surprisingly the link between ideology and social movement behavior has been overlooked in recent research (as exceptions see Melucci 1989; Downey 1986; Perrow 1979; Snow and Benford 1988; Scott 1990; and Klandermans 1991). Many early collective behavior theorists viewed ideology as a source of irrational political action and discounted it on these grounds (e.g., Smelser 1963); more recently, the RM approach has treated ideology as irrelevant. When RM theorists discuss variations in the structure and tactics of social movement organizations, these patterns are attributed to differences in the resource environment or other nonpolitical factors (e.g., Zald and Ash 1966). Even the theorists who explore the supposedly new aspects of NSMs often do so in a descriptive manner, without clearly tracing the source of these changes to ideological factors.

We see European environmental groups as following a pattern of behavior in which the ideology of an organization interacts with its resource needs and opportunities. This is not a rejection of the RM approach, but rather an elaboration of the approach—where the ideological identity of an organization guides its choices of what goals to pursue, what tactics are appropriate to its position, and what resources are available for mobilization. Viewing the environmental movement, or any social movement, in this way enables one to comprehend more clearly the actions of various environmental groups and ultimately to understand the larger social and political implications of these movements for advanced industrial societies.

A Focus on Environmental Interest Groups

RM theorists often speak of a social movement industry to portray the network of individuals and organizations that comprise a social movement. In the broadest terms, McCarthy and Zald (1977, 1217) define a social movement as "a set of opinions and beliefs in a population that represents preferences for changing some elements of the social structure and/or the reward distribution of a society." At this level, the environmental movement includes all those individuals who are concerned about pollution, nuclear power, the quality of life, and other related issues. Even though most of these individuals have no formal ties to an environmental organization, their views are nonetheless an important element of the movement. Public opinion surveys that document widespread popular support for environmental legislation help to legitimize the movement and its activities and to spur organizations into action, even if most of these citizens remain supporters only in conscience.

At another level, one might define the movement as the smaller group of individuals who are personally active in green organizations. Collectively, these citizens provide vital financial and personal resources for environmental action. Their political beliefs and expectations help define the identity of the green movement, and these orientations in turn structure their behavior in pursuit of environmental causes (see, e.g., the studies of environmental activists conducted by Cotgrove 1982; Milbrath 1984; and Diani 1988). Research on movement activists can answer important questions about the mobilization processes and political identity of a movement as shaped by these supporters (Klandermans et al. 1988).

The environmental movement industry is also composed of a variety of political organizations. The simplest organizations are the citizen action groups concerned with the problems of a specific locale. Such local groups often helped to develop initial public awareness of environmental problems and first mobilized participation in environmental actions. In addition, virtually all Western democracies contain several larger regional or national public interest groups concerned with green issues. In many European states, membership in such voluntary organizations exceeds formal membership in other political organizations—including political parties. Political parties themselves are another potential component of the environmental industry. The parties normally bear responsibility for the representation and articulation of issue interests within the European system of government. The environmental issue has stimulated new political demands for established parties and has led to the creation of new green parties in many Western party systems. RM theorists also maintain that external organizations that provide support for the environmental movement—such as other special interest associations, political groups, and public or private agencies—constitute part of the environmental industry.

All of these elements are important to the environmental movement, and choosing a level of analysis—from public supporters to environmental parties—influences both the relevant research questions and the conclusions one draws about the movement. A study of national environmental groups will, almost by definition, find greater evidence of institutionalization and bureaucratization than will studies of local citizen groups. Environmental attitudes and beliefs inevitably appear more diffuse and imprecise in surveys of the general public than in comparable studies of environmental activists. Yet the environmental movement comprises all of these elements.

In spite of these multiple elements, this study argues that environmental interest groups may provide the key to understanding the fundamental nature of contemporary environmentalism. Although many Europeans are concerned about environmental issues, they must coalesce into a political force if their views are to affect society and political decision makers. Because of the present structure of the movement, environmental interest groups represent the primary institutional base of the movement in most nations. In contrast to local citizen action groups, environmental interest groups are ongoing, institutionalized advocates for political action that reach beyond the concerns of a specific locale. Political parties aggregate many political demands, whereas environmental interest groups are solely advocates of one movement.

As RM theory maintains, environmental interest groups are influential in channeling public support for environmental issues. These groups translate public interest into a potential political force and mobilize popular support behind this agenda. When Greenpeace mounts a television campaign or the Italian League for the Protection of Birds asks members to support bird legislation, both groups are representing public interests and putting pressure on policy makers. Environmental groups can serve as political reference points—similar to endorsements by unions or business associations—letting the public know which policies are important and which strategies should be adopted (Pierce et al. 1992).

As we shall see, environmental interest groups are also important participants in the political process of Western democracies. Policy influence in contemporary societies requires the formal representation of interests. Legislation must be monitored, close contact must be maintained with the bureaucracy, authoritative witnesses must be available to present environmental views, and the organization must be able to educate and mobilize its supporters. Because of their institutional base, environmental groups are better positioned than ad hoc citizen groups to carry out these activities. In pursuing these activities, environmental groups decide on the agenda and political direction of the movement. In short, it is our contention that environmental interest groups provide a key to understanding both the environmental movement and the broader political changes occurring in advanced industrial democracies.

Because of our interest in the environmental movement as a new political participant in advanced industrial societies, we chose to examine environmental groups comparatively in several Western democracies. The study includes the ten member states of the European Community (EC) as of December 1985: Belgium, Britain, Denmark, France, Greece, Holland, Ireland, Italy, Luxembourg, and West Germany.[13] We selected the EC because these nations have common interests and a growing international involvement on environmental issues. The Community issued its Fourth Environmental Action Plan in 1986, and the Single European Act of 1987 modified the Treaty of Rome to give the Community new legislative authority on environmental policy. Choosing the EC involved pragmatic considerations as well. The Directorate General for the Environment (DG XI of the EC) publishes comparative data on environmental conditions in the EC member nations (Commission of the European Communities 1990). In addition, the semiannual Eurobarometer surveys provide extensive information on public attitudes toward environmental issues in all EC nations. Fortuitously, we were able to coordinate our interviewing of European environmental groups with a special Eurobarometer on environmental attitudes that the EC conducted in spring 1986 (Eurobarometer 25).[14] This afforded a unique opportunity to link public opinion to the findings of our environmental group study.

We used four criteria in selecting environmental organizations within these countries (see also appendix). First, we focused on national organizations that exert influence on a national level, thereby excluding the plethora of local citizen action groups and initiatives within the environmental movement—though these local groups often maintain formal or informal links to national organizations. Second, we excluded single-issue groups, such as those concerned only with nuclear power, noise pollution, or transportation policy. These specific interests are represented within the groups we chose, but we wished to emphasize broader environmental interests. Third, we selected groups to reflect the variety of ideological and organizational forms within the movement. In each nation we consciously sought out the major representative of the new wave of the environmental action, as well as the national bird societies and other major conservation groups. We included both mass-membership and nonmembership organizations. Fourth, we selected groups with a somewhat permanent organizational structure; this criterion was often implicit, because well-established groups were the easiest to identify and contact. We felt that groups that maintain an ongoing organizational presence are more likely to influence the evolution of the movement and national environmental policy.

Using these criteria, we selected the relevant environmental organizations in each nation. Research on the environmental movement frequently focuses on unconventional and antiestablishment groups: Greenpeace, Friends of the Earth, and other New Left groups. These groups are generally identified as fitting the pattern of New Social Movement theory. In addition, our definition of the environmental movement includes conservation and wildlife protection groups. Most important from a theoretical perspective, conservation groups generally lack a challenging ideology and thus come closer to the value-free behavior of RM theory. Thus the contrast between these two types of environmental groups is an essential element in testing our model of ISA.

The inclusion of both types of environmental groups also enables us to examine political relations within the movement. These organizations share many of the same environmental interests, as well as memberships and a financial base that partially overlap. During our interviews we uncovered an extensive network of relations across the environmental spectrum. In several nations, there are national umbrella organizations in which group leaders meet on a regular basis. Group representatives repeatedly mentioned instances where groups decided, formally or informally, on a division of responsibilities or coordinated their separate activities. Sometimes, a group planned its own activities to complement the efforts of other groups. Extensive personnel and social networks further interconnect these groups.

During the winter of 1985–86 our research team traveled to Europe and conducted in-depth interviews with representatives of sixty-nine environmental organizations (see chap. 4 and appendix). This set of organizations comes close to including the universe of all important national environmental organizations in Western Europe. Equally important, this study includes the ideological and political diversity of the European environmental movement, ranging from traditional wildlife protection groups to critics of advanced industrial societies, from amenity societies to *Alternativbewegungen*. The study includes forty-eight mass-membership groups, twelve national umbrella organizations, six research or educational institutions, and three "by invitation only" groups of environmental elites.

We conducted personal interviews with a top-level official of each organization, whether the director, the information officer, or another senior official.[15] The interviews focused on perceptions of environmental conditions, the organizational structure of the group, its policy activities, and the political views of group representatives. These interviews provide the central resource for this study, supplemented by documentary materials collected from each organization, additional interviews conducted with environmental experts in each nation, and secondary analyses of other published sources.

The environmental movement is a dynamic force that has continued to evolve since our data were collected. We should therefore avoid a static view of the movement; any picture of environmental interest groups reveals the movement only at a particular time, regardless of how rich the tones and how precise the detail of the research. Still, this book provides a detailed and comparative picture of environmental action in Western Europe heretofore unavailable. This research can also speak to the future potential of the movement.

Plan of the Book

The next two chapters locate contemporary environmental groups in a larger historical and political context. Chapter 2 traces the historical evolution of the environmental movement from its beginnings in the late 1800s to the present. The chapter describes both the evolution of specific environmental organizations and their respective environmental ideologies. Chapter 3 examines the breadth and depth of public support for the movement and the policies it advocates and considers several possible explanations for the public's interest and broad enthusiasm for environmental action.

The second section of the book focuses on the organizational characteristics of the movement. Chapter 4 describes the basic features of the environmental organizations included in this study: their origins, resources, and political orientations. We also examine the internal structure of environmental interest groups and assess the impact of a group's ideology on its organizational structure and decision-making processes. Chapter 5 investigates the environmental elites we surveyed; we describe their social background, the paths that led them toward their present position, and their personal political values.

The next set of chapters examine environmental groups as political actors. We begin in chapter 6 by describing the environmental issue agenda from the perspective of the movement: what issues are most important, what steps must be taken to address these issues, and what are the perceived impediments to environmental reform? Chapter 7 looks at the alliance patterns of environmental interest groups: Which social and political groups are seen as the friends and foes of the movement, and to what extent is their cooperation among groups within the movement? Chapter 8 focuses on the various strategies and tactics that groups employ in pursuit of their cause and the role of ideology and other organizational characteristics in guiding group decisions about what tactics are necessary and appropriate. Because of the centrality of political parties in the European governmental structure, chapter 9 examines the link between environmental groups and political parties in closer detail.

Chapter 10 reviews the evidence in support of our framework of Ideologically Structured Action and, on this basis, discusses the potential implications of the environmental movement for Western European democracies.

The environmental movement is, in a sense, a critical case study. We see environmentalism as illustrating the forces of social change now developing in advanced industrial societies. At the same time, at a theoretical level the environmental movement stands at the nexus of several fundamental research questions about the nature of public issue interests, the structure of political organizations, and the changing style of citizen politics in these societies. Certainly, the long-run consequences, and the even more immediate impact, of the environmental movement in changing public priorities and styles of action remain uncertain. Yet our study is well-positioned to look at the development of environmentalism at a crucial period in its evolution, as it—along with other New Social Movements—attempts to become a major force for social and political change in advanced industrial democracies.

2

MOVEMENTS IN CONTEXT

Chapter Two

THE EVOLUTION OF

ENVIRONMENTALISM

The scenario is all too common in contemporary industrial societies: Amsterdam's garbage was exceeding the disposal capacity of the city, and a new dumping site had to be found. City leaders wanted to dump the refuse in the Naarder Sea, just east of Amsterdam, but in this case members of the Amsterdam social elite protested the loss of a bird-breeding area that was home to the rare spoonbill, the purple heron, and other wildlife. They formed a citizens' group to protest the city's actions and lobbied government officials for protection of the area. The following year the group purchased the threatened land and converted it into a nature preserve. Instead of a refuse dump, the Naarder Sea became a wildlife sanctuary and a national treasure of Holland.

Even though citizen action groups are now commonplace, this was an unusual event because of its timing. It did not happen during the past decade, or even during the rapid growth of environmental groups in the late 1960s and early 1970s. This citizen protest occurred in 1904 and led to the formation of the Association for the Preservation of Natural Monuments (Vereniging tot behoud van Natuurmonumenten), which is now the largest conservation group in Holland.

Analysts often equate the environmental issue with the "new" political controversies of advanced industrial societies, but the origins of the movement date back at least to the 1800s. The period 1880–1910 included the first major wave of environmental action in Western Europe. Citizens in several nations formed new voluntary groups to protect wildlife, preserve natural areas of national significance, and conserve nature. Many of the first major pieces of environmental legislation and land-use planning also date from this era. After a long period of relative dormancy, a second wave of environmentalism swept across Europe beginning in the 1970s. New environmental groups emerged to deal with current problems and press their concerns onto the political agenda. In many areas, these new groups moved beyond the issue interests and social base of traditional conservation groups, but the basic interests of the old and new groups often overlapped and their efforts frequently reinforced one another.

Analysts of the environmental movement continue to debate the nature of the relationship between older conservation groups and the modern environmental movement.[1] Karl Werner Brand (1990) sees both waves of environmentalism as similar phases in an ongoing cycle of modernization in Western societies. Brand maintains that periods of rapid industrialization and modernization stimulate feelings of cultural criticism by the public, which leads to the formation of environmental groups and other alternative movements. Sidney Tarrow (1983), Wilhelm Bürklin (1987a), and others offer similar cyclical explanations of social mobilization, and Seymour Lipset has discussed the broader evolution of Western democracies in similar terms of recurring waves of social reform and counterreaction (Lipset 1981; see also Huntington 1981). In contrast, analysts of New Social Movements (NSMs) hold that differences in ideological orientation, political style, and social base sharply distinguish traditional conservation groups from the contemporary ecological movement (Offe 1985; Cotgrove 1982; Milbrath 1984; Duyvendak and van Huizen 1982).

In this chapter we examine the historical evolution of the environmental movement from its initial wave of mobilization in the late 1800s through the emergence of contemporary environmental groups. The first objective is to recount the historical record of environmental action. The more basic goal is to use this historical record to learn about the factors that led to these mobilization waves and from this to determine the similarities and differences between various environmental organizations.

The Beginnings of Nature Conservation

The modern nature conservation movement began to develop in Western Europe during the latter half of the nineteenth century, though an exact date is difficult to pinpoint.[2] Several factors produced the first stirrings of the movement. In most nations, the environmental consequences of the Industrial Revolution were becoming manifest during this period. Urbanization and industrialization had transformed landscapes, and the harmful effects of these processes were destroying wildlife and natural areas, as well as polluting the environment. Historical accounts of the period stress the scale of industrial pollution, the impact of railroad expansion, the construction of public sanitation facilities, and similar developments as factors that focused public attention on environmental matters.

The growth of the natural sciences during that century also stimulated awareness of environmental problems. Biologists and botanists studied and catalogued the natural environment, documented the loss of habitats and species, and traced these problems to industrialization. The German botanist Ernst Häckel first used the term *ecology* in 1866. The natural sciences developed a formal standing in Europe during the latter half of the century, and the growing membership of natural history societies provided a popular basis for environmental action.

These changes in the objective circumstances of life were joined by an even more important shift in the Zeitgeist of the upper class in most European societies. European intellectuals increasingly challenged the belief in rationalism and progress spawned by the Enlightenment. In France, a romanticist sentiment developed by mid-century (Duclos and Smadja 1985; Vadrot 1978). Authors such as Chateaubriand, Victor Hugo, and Alfred de Vigny reflected the rural and pastoral tradition (*la tradition champêtre*) of Jean-Jacques Rousseau and portrayed an idealized view of nature. Similar romantic notions of the natural order and a longing for a return to nature developed in Belgium and Holland (Tellegen 1981). In Britain, Victorian romanticism replaced the Darwinist view of nature identified with laissez-faire political economists (Weiner 1981; Lowe 1983; Thomas 1983). Instead of dominating nature, Victorians saw the protection of nature and wildlife as a measure of humankind's higher order. Thus, some of the leading British philosophers of the period—John Stuart Mill, John Ruskin, and Lord Avebury—were founding members of conservation groups. The idealization of nature and the establishment of communes were visible parts of the British counterculture of the late 1800s (Marsh 1982; Weiner 1981). By the end of the century undercurrents of romantic sentiment also ran through German society, perhaps best typified in the *Wandervogel* movement that enabled upper-class urban youth to relish the joys of nature through weekend country hikes and camping expeditions. The last decade of the century was a period of general cultural criticism in Germany (*Zivilisationskritik*) that idealized nature and the goal of returning to the land (Linse 1983; Sieferle 1984). Only Southern Europe, which had not yet experienced the full force of industrialization and modernization, did not share in this critique of modernity.

These romantic currents within Europe were both a reaction to the tremendous social changes that these societies were experiencing (an early version of "future shock") and a criticism of the direction they were taking. The economic depression of the late 1800s further heightened these anti-industrial sentiments. An idealized view of nature and the natural order provided a source of stability and reassurance in a rapidly changing world; these sentiments also stimulated actions to protect (or create) this image of the natural order.

European nations dealt with these forces in different ways. In some cases, objective conditions and national economic needs prompted the first conservation activities. In others, moral and aesthetic values seemed more apparent in stimulating action. In almost all of Northern Europe, however, the common result was the proliferation of conservation groups and public pressure for conservation legislation throughout the late 1800s and early 1900s.

One component of the conservation movement, and probably the largest, involved the protection of birds. The first wildlife campaign in Britain, for example, sought to protect seabirds from wanton slaughter by sportsmen (Sheail 1976; Lowe 1983). As birds nested along the Yorkshire coast, hunters had traditionally used them as live targets, killing hundreds a day and leaving the bodies where they fell. Led by the local clergy, residents of the area protested this destruction and in 1867 formed a society for the protection of seabirds. This citizens' lobby used a media campaign in the *Times* of London, aid from the Royal Society for the Prevention of Cruelty to Animals, and scientific support from the British Ornithological Union to pressure their MP into introducing a bill protecting seabirds, which Parliament passed in 1869. The Yorkshire residents pursued this action for basically altruistic reasons, namely, to avoid unnecessary harm to wildlife and to protect whole species of seabirds from possible extinction.

Concern over bird protection was spreading on the European continent (Hayden 1942). In 1868, a group of German farmers and foresters petitioned the Austro-Hungarian foreign office to protect migratory birds that were beneficial to German agriculture. This petition led to a series of international conferences (Budapest in 1881, Vienna in 1884, Paris in 1895) that debated the issue and discussed possible solutions. Finally, the participating nations reached an agreement in 1902 that specified a list of "beneficial" birds to be protected by the countries signing.[3]

A further stimulus for the bird protection movement was the growing use of feathers and other forms of bird plumage in women's dress fashions (Doughty 1975). Around the turn of the century, the dictates of fashion created a large international market in bird feathers, centered in Paris, London, and New York. Both exotic species, such as the osprey and bird of paradise, as well as common birds, such as the robin and hummingbird, were sacrificed to ladies' fashions; egret plumes decorated elaborate headwear or a robin's breast served as a lapel ornament. Bird preservationists claimed that at the beginning of the twentieth century hunters were annually killing more than two hundred million birds for their feathers.

This wanton destruction of wildlife for nonessential reasons violated the humanist and Victorian sensibilities of many upper-class women in Europe. The consequences of these sentiments were perhaps most obvious in Britain, where reactions to the plumage trade led to the proliferation of bird protection groups (Doughty 1975; Sheail 1976). The Plumage League and the Selbourne Society were formed in 1885 as separate groups opposed to the feather trade; shortly thereafter they merged under the Selbourne Society title. In 1889, a group of women in Manchester who had been unable to join the all-male British Ornithological Union established the Society for the Protection of Birds (SPB). In 1891 the society developed a broader geographic base when it merged with another local women's group, the Fur, Fin and Feather Folk of Croydon; by 1893 the SPB boasted a national membership of nine thousand. The Selbourne Society and the SPB often differed in tactics and competed for political influence, but both groups pressed for an end to the exploitation of birds for fashion. The SPB developed its social and political contacts more rapidly and received a royal charter in 1904 to become the Royal Society for the Protection of Birds (RSPB). The dramatic and emotional appeals of the RSPB on behalf of bird protection were very effective in mobilizing public opposition to the killing of birds for their plumage, more so than the dry scientific studies of the British Ornithological Union. Newspaper stories depicting the destruction and barbarity that accompanied massive bird kills and public letter writing campaigns preached that feather fashions violated the Victorian norms of compassion and refinement. Even so, bird protection groups faced several decades of lobbying until the plumage trade in Britain was effectively halted by proclamation during World War I and then by legislation in 1921.

The public discussion of the plumage trade diffused public interest in the bird protection issue throughout Europe. The Dutch Association for Bird Protection (Vereniging tot Bescherming van Vogels) and the German Federation for Bird Protection (Deutscher Bund für Vogelschutz, or DBV), were both established in 1899; about the same time, the Irish Society for Bird Protection was formed. These groups rapidly gained public acceptance; the DBV, for instance, grew from six thousand members in 1902 to forty-one thousand in 1914 (Dominick 1992, 53). Denmark passed its first bird protection legislation in 1894, and this was followed by the establishment of the Danish Ornithological Society (Dansk Ornitologisk Forening, or DOF) in 1906. With the creation of the French League for Bird Protection (Ligue pour la Protection des Oiseaux, or LPO) in 1912, a national lobby for bird protection existed in nearly every Northern European nation. Although these groups were almost exclusively concerned with protecting birds, ending the plumage trade, and related issues, their activities contributed to the general spirit of the times.[4]

Preservationist groups constituted a second component of the early conservation movement. A variety of organizations emerged throughout Europe to preserve their nation's historical sites and areas of natural or environmental significance. In Britain, efforts to protect open spaces and common lands eventually led to the formation of the National Trust in 1895. The Trust acted as a holding company for those wishing to donate land for preservation as part of Britain's national heritage. Parliament recognized the Trust as a unique statutory body in 1907 and charged it with permanent preservation of property for the benefit of the nation. This legal status gave the Trust the ability to declare that donated lands would be held in perpetuity and could be divested only with parliamentary approval. Besides acting as a land-holding company, the Trust also formed a voluntary association that pressured policy makers on conservation issues. The National Trust initially focused its attentions on the protection of historical sites rather than areas of natural beauty or environmental importance. In 1912, the Society for the Promotion of Nature Reserves (SPNR) was formed to work with the National Trust in developing nature reserves that would both protect wildlife and preserve areas in their natural state (Sheail 1976). Although not as successful as the National Trust in acquiring land or financial support, the SPNR laid the groundwork for future conservation efforts in Britain.

Germany was another early leader in the preservationist movement, but the German experience differed from the rest of Europe in two important ways. First, government assumed responsibility for preservation of significant national sites. Conservation was the responsibility of the state governments, and many states actively worked to protect areas of historical or environmental importance. In 1906, for example, Prussia created an office for the protection of national monuments, the forerunner of the modern environmental ministry, and appointed an internationally renowned conservationist, Hugo Conwentz, to the office. Second, the federal structure of policy making led to the regionalization of the conservation movement. There were few equivalents to the national organizations in Britain and other European democracies; instead, conservation organizations tended to form at the state level. Conservationists formed the German Federation for Protection of the Homeland (Deutscher Bund Heimatschutz) in Dresden in 1904, the Nature Protection Federation of Bavaria (Bund Naturschutz Bayern) in 1913, and the Nature Protection Ring (Naturschutzring Berlin-Brandenburg) in Berlin. Groups also formed in other states and at the local level. Preservation of the nation's heritage became a measure of German national patriotism, leading to the early creation of nature preserves and the passage of conservation legislation (Dominick 1992; Conwentz 1909; Wey 1982).

Comparable preservationist groups developed in most other Northern European nations. The Association for the Preservation of Natural Monuments, mentioned previously, was established to protect the Naarder Sea; the organization continued to develop nature preserves throughout Holland. It now manages roughly 200 reserves amounting to more than 120,000 acres. In 1911, conservationists formed another organization, Heemschut, to protect cultural monuments in Holland, such as old buildings and historical sites. Around the turn of the century, French and Belgian citizens became interested in protecting places of cultural significance and natural beauty, creating the Société pour la Protection des Paysages and the Société Nationale pour la Protection des Sites, respectively. In 1911 Danish environmentalists established the forerunner of the Danish Association for Nature Conservation (Danmarks Naturfredningsforening, or DN) in 1911; in 1925 it became a national organization promoting the conservation of nature and public education on conservation issues.

Although many preservationist groups were interested in the protection of wildlife, a third distinct component of the conservation movement focused exclusively on the protection of nonavian species. Indeed, the oldest of the major European conservation groups is the French National Society for the Protection of Nature (Société Nationale de Protection de la Nature, SNPN). Originally formed in 1854 as the Société Impériale Zoologique d'Acclimation, it was an association of zoologists who wanted to protect the wildlife of France and its empire, both through legislation and by establishing nature preserves. A similar organization, the Society for the Preservation of Wild Fauna of the Empire (SPWFE), was formed in Britain in 1903 (Fitter and Scott 1978).[5] Conservationists created the society to oppose the efforts of a colonial governor to abandon a game reserve in the Sudan. The early leadership of the SPWFE described itself as "a modest and unpretentious group of gentlemen of wide experience of the outposts of the Empire and a common enthusiasm for the preservation from destruction of many of its fauna (8)."

Despite the commendable goals of these two wildlife organizations, their ultimate aim was not the protection of species from human assault but the preservation of species for hunting and other uses. The French society, for instance, celebrated wildlife preservation at their annual meeting with a dinner of exotic game; one could hardly imagine a contemporary conservation group performing the same ritual. Similarly, the British society was largely composed of royalty and former colonial administrators who had enjoyed the safari experience and wanted to create big-game preserves, not wildlife sanctuaries. Critics of the SPWFE lampooned it as a group of "penitent butchers," a nickname that stayed with the group because its members seemed to develop a concern for wildlife only after they had finished hunting.

Although we can distinguish between separate components of the conservation movement, in practice their actions were often complementary. Just as the National Trust and SPNR worked together on historical and landscape preservation in Britain, the Association for the Preservation of Natural Monuments and Heemschut cooperated in the same way in the Netherlands. National societies for the prevention of cruelty to animals (pets and domesticated animals) often facilitated the development of bird and wildlife preservation associations, such as the RSPB in Britain and the Association for Nature Conservation in Denmark. Wealthy benefactors and board members overlapped among groups, and the voluntary associations often drew upon the same popular base of upper-class European society. Thus, together these separate organizations combined to produce a general mobilization period for conservation issues.

Almost as quickly as the conservation movement grew in the early 1900s, its momentum began to slacken a short time later, in part because of its own success. National legislation and international agreements on bird protection had at least partially addressed the problems that initially led to the formation of the bird protection lobby. Conservationists also made progress on wildlife preservation. By the 1910s, there were organizations aimed at the preservation of historical landmarks and the natural environment in most Northern European nations. In short, policy success partially obviated the rationale for additional political efforts. World War I and its consequences further undercut the growing momentum on conservation issues. In the years before the war, European conservation organizations were developing an international agenda for the protection of nature in Europe and its colonies (Boardman 1981). These efforts led to the establishment of a Commission for the International Protection of Nature and plans for a series of international congresses. The war broke out before the commission really began to function. The war similarly sidetracked international cooperation among bird protection associations. World War I not only undermined the groundwork for international cooperation, it also shifted public attention from conservation issues to matters of greater immediacy: postwar reconstruction. Historical accounts of the SNPR, for example, note that the organization stagnated because it was formed just two years before the outbreak of World War I.

The interwar years were a period of relative dormancy for the conservation movement. Environmental problems certainly had not ended, and a series of environmental problems surfaced during this period (such as oil pollution at sea and newly endangered species). Public interest in conservation issues dwindled, however. Postwar reconstruction and then the Great Depression focused public interest on more immediate economic difficulties. Although most of the major national conservation groups discussed above continued to exist, their membership roles stabilized at fairly modest levels.[6] In fact, only one of the major contemporary environmental groups in Europe, the Council for the Protection of Rural England (CPRE), was established during the interwar period (1926). At the international level, agreements were reached on pollution from shipping, the protection of whales, and the protection of African wildlife. Lynton Caldwell (1984) concludes, however, that these agreements represented very small gains, and later events showed that their provisions were inadequate or soon overtaken by new environmental threats. The international momentum that had been slowed by World War I was never reestablished. In summary, the mobilization wave of 1880–1910 had raised the European environmental movement to a plateau and created the organizational infrastructure necessary to remain at this level, but the interwar years generally represented a continuation along this plateau.

The Modern Revival of Conservationism

The conservation movement began to revive following World War II. In Britain, wartime destruction created new opportunities for conservationists. Britain's wartime needs for agricultural products had temporarily transformed the landscape of rural England; conservationists saw postwar planning and reconstruction as an opportunity to preserve what remained of the natural British countryside. In addition, popular sentiments toward government had changed as a result of the war; Britons attributed greater responsibility to the state in addressing social needs. One illustration of this new sentiment was the passage of the Town and Country Planning Act in 1947, which granted the state control over planning for all land in Britain. Legislation in 1949 established the Nature Conservancy as a government agency charged with establishing a national system of publicly owned nature reserves.

This same conservation revival occurred in other Western European nations. Dutch, Belgian, and French conservation movements rapidly reorganized after the war. As in Britain, these groups were motivated by a reaction to the destructive impact of war on the environment and by the need for environmental planning in postwar reconstruction. In addition, conservation concerns spread to new areas: the modern Italian conservation movement began in 1951 with a protest by a group of writers and artists opposed to the urban development projects that threatened the destruction of historical areas in the center of Rome. These efforts led conservationists to establish Italia Nostra in 1955, an organization dedicated to the preservation of the cultural heritage in Italy's architecture, museums, and rural areas. Perhaps the greatest change came in postwar West Germany, where conservation and preservation groups were refounded as part of the new democratic system. Older groups, such as the German Federation for Bird Protection and the Nature Protection Federation of Bavaria had to sweep away the legacy of the Third Reich and begin rebuilding their organizations virtually from the foundations.[7] In addition, new conservation groups formed; conservationists created the German Nature Protection Ring (Deutscher Naturschutzring, or DNR) in 1950 as an umbrella group to unite many of these diverse organizations within a single conservation lobby. The DNR now boasts a membership of 95 organizations with a combined individual membership of 3.3 million.

Another impetus for the postwar conservation movement came at the international level (Boardman 1981; Caldwell 1984). The Swiss League for the Protection of Nature sponsored a conference in 1946 that assembled the leading conservationists in Europe, such as P. G. van Tienhoven from Holland and Julian Huxley from Britain. This meeting helped to reestablish old networks of international support and cooperation. Under the auspices of UNESCO, European conservationists worked toward the development of a new international conservation body. In 1948, a charter established the International Union for the Conservation of Nature (IUCN) as an international association of government conservation offices and nongovernmental conservation organizations.[8]

With many conservation and environmental issues taking on international dimensions, the IUCN became a central participant in the international conservation network. The IUCN, however, was largely an intergovernmental body that facilitated the exchange of scientific information and policy advice between governments and other international organizations. These administrative functions limited its ability to mobilize public opinion and financial support on behalf of conservation causes. The IUCN's limitations (and its tenuous financial situation) led European conservationists to create a new organization designed to involve the public in conservation issues: the World Wildlife Fund (WWF). Conservationists established a national branch in Britain (1961), and WWF affiliates soon followed in France (1963), Germany (1963), Belgium (1965), Denmark (1965), Italy (1966), and the Netherlands (1972). WWF became the public representative of the international conservation movement, the first multinational environmental group, and it dramatically expanded the financial and popular support for international conservation issues (Boardman 1981, 108–10; Nicholson 1970).

Despite this progress, the postwar years were a difficult period of rebuilding for the conservation movement. Membership in conservation groups grew throughout most of Europe, but only slowly, and financial resources remained limited. The success of the movement was measured not by its accomplishments—though new legislation was being passed and membership was growing—but by its efforts to restore environmental issues to the public agenda. The greatest accomplishment was the formation of an organizational base that was to support a dramatic new surge in the movement.

A New Environmental Wave

In the late 1960s, environmental issues again became a salient political topic in Europe.[9] Public interest in conservation and environmental issues surged. In addition, a host of new environmental organizations emerged within a few years, often attracting widespread support from new sectors of society. Although the public's growing environmental consciousness often remained centered on unresolved conservation problems, new environmental problems experienced by advanced industrial democracies began to emerge: nuclear power, resource shortages, toxic waste, acid rain, and protection of the quality of life.

This new wave of environmental mobilization differed in important ways from the earlier conservation movement; but like the earlier wave, it resulted from the convergence of several factors, one of which was the scientific and educational network created by conservation organizations. This network initially provided the information that alerted citizens and public officials to the world's growing environmental problems. The clearest example is the writing of Rachel Carson, a former biologist at the U.S. Fish and Wildlife Service. Her book, *The Silent Spring* (1962), sounded the warning of an impending environmental crisis and had an enormous impact on both sides of the Atlantic. Both European and American authors followed with a host of other books that stimulated this growing environmental consciousness: Stuart Udall, *The Quiet Crisis* (1963); Jean Dorst, *Avant que nature meure*

(Before Nature Dies, 1965); Rolf Edberg, *Spillran av ett moln* (On the Shred of a Cloud, 1966); Barry Commoner, *Science and Survival* (1966); Paul Ehrlich, *The Population Bomb* (1968); Gunther Schwab, *Der Tanz mit dem Teufel* (Dance with the Devil, 1958); Bodo Manstein, *Im Würgegriff des Fortschritts* (The Stranglehold of Progress, 1961); and Howard Odum, *Environment, Power and Society* (1971). Few of the issues raised by these authors were new. The problems of indiscriminate use of DDT, for example, were well known by specialists in advance of Carson's book; but governments were slow to act until *The Silent Spring* awakened the public to the menace. The primary impact of these books was to draw public attention to these growing problems.

The writings of conservationists took on added force because of several dramatic environmental crises that occurred in the later 1960s. In 1967 the supertanker *Torrey Canyon* ran aground in the English Channel and spilled tens of thousands of tons of crude oil into the sea. This enormous spill fouled the coasts of Cornwall and Britanny, destroying birds and aquatic life and causing millions of dollars in damages. Two years later, toxins leaking into the Rhine River produced a massive fish kill. Rotting, poisoned fish covered the banks of the river, a main source of drinking water for millions of Europeans. The blowout of an oil drilling platform off the Santa Barbara coast in early 1969, and the subsequent destruction of the beautiful California coastline, produced reverberations that reached all the way to Europe. Then late in 1969, thousands of dead birds mysteriously washed up on the banks of the Irish Sea, generating front-page headlines in the London newspapers and governmental inquiries that eventually attributed the problem to the side effects of industrial pollution. Environmental problems were no longer something that affected only bird-watchers or rural preservationists; the potential threat to even the average individual was becoming obvious. Moreover, these spectacular events stimulated greater public awareness of other environmental problems within a nation or local area, which further heightened popular demands for new environmental protections.

For at least some parts of the modern movement, the growth of environmentalism also reflected a cultural criticism of advanced industrial societies that arose from Europe's cultural ferment in the late 1960s (a new form of *Zivilisationskritik*). European youth attacked the insatiable materialism and excess consumerism of advanced industrial societies—and environmental problems illustrated their radical commentary. For example, scholars trace the origins of the modern French environmental movement and anti-nuclear-power movement to the forces spawned by the May Revolts of 1968 (Vadrot 1978; Chafer 1982). The revolts gave environmentalism an ideological theme:

the deterioration of the environment reflected the inevitable consequences of unbridled economic growth and the overcentralized French state. Similarly, the social criticisms of the Provo movement (1965–67) and the Kabouter movement (1968–70) provided a political and ideological starting point for radical environmental action in Holland (Tellegen 1981, 1983; Duyvendak and van Huizen 1982; Jamison et al. 1990, chap. 3; von Moltke and Visser 1985). Dutch youth cited the problems of pollution and environmental protection as examples of the failures of the economy and the bureaucratic elitism of Holland's consociational democracy. The formative influence of the student movement and the protests of the late 1960s are equally prominent in accounts of the beginnings of the ecology movement in West Germany and Denmark (Brand et al. 1986; Kitschelt 1989, chap. 4). In Belgium, youth protests for regional decentralization and democratization of society gradually broadened their scope to include environmental issues. In addition, Flemish environmentalism developed as an offshoot of a charismatic Catholic movement, Live Differently (Anders Gaan Leven), a movement that encouraged its adherents to adopt a life-style of voluntary simplicity and reject the dominant social paradigm of industrial society (Kitschelt and Hellemans 1990, chap. 2).

The overlap between the student movement and environmental movement was also important in defining the activists and supporters of green causes. The student movement spawned a new generation of assertive young political activists who provided a leadership cadre and activist core for many of the newly forming environmental groups. Similarly, participants in the new wave of environmental action were heavily drawn from the university graduates of the 1960s and 1970s. The children of Europe's postwar economic miracle provided a base for modern environmentalism.

The public discussion of conservation and environmental issues was further stimulated by governmental actions. The Council of Europe declared 1970 "European Conservation Year" to stimulate public awareness. Even more important was the convocation of the United Nations Conference on the Human Environment held at Stockholm in 1972, a culmination of several years of international meetings and environmental studies (Caldwell 1984). The Stockholm conference assembled officials from 114 nations, scientific experts, and an exceptionally large number of conservationists and environmentalists. It produced specific proposals for governmental action in the form of a United Nations Environmental Program (UNEP). The primary accomplishment of the Stockholm conference, however, was that "it legitimatized environmental policy as a universal concern among nations, and so created a place for environmental issues on many national agendas where they had been previously unrecognized" (Nicholson 1970, 110; see also Caldwell 1984, chap. 3).

New books and television productions on environmental topics also spurred action on environmental issues. The most influential publication was *The Limits to Growth,* published by the Club of Rome, which contained dire forecasts for humankind and the environment (Meadows et al. 1972). This, along with the Stockholm meetings, stimulated many public figures to accept environmentalism as a significant issue. A multitude of other books aroused public interest with more popular accounts of the problem, such as J. E. Lovelock, *Gaia: A New Look at Life on Earth* (1979); Edward Gold-smith et al., *Blueprint for Survival* (1972); Erhard Eppler, *Ende oder Wende* (End or Change, 1975); Herbert Gruhl, *Ein Planet wird geplündert* (A Planet Is Plundered, 1975); Laura Conti, *Cos'è l'ecologia?* (What Is Ecology? 1977); Dario Paccino, *L'imbroglio ecologico* (The Ecological Deception, 1972); and André Gorz, *Ecologie et liberté* (Ecology and Liberty, 1977).

In the years immediately before and after the Stockholm conference, many European governments took up the challenge of addressing the environmental problems of advanced industrial societies. In 1970, Georges Pompidou introduced a host of measures in response to the growing demand for environmental protection in France; the government followed this action with the creation of the Ministry of the Environment in January 1971 (Vadrot 1978, chap. 1). The West German government strengthened its commitment to environmental protection in the wake of the Stockholm conference; it revised the Basic Law in 1972 to grant the federal government jurisdiction on environmental matters and then implemented twenty-six major pieces of legislation during the 1972–76 session of parliament—more environmental legislation than was passed in the first two decades of the Federal Republic (Klingemann 1985). The establishment of a Danish Ministry for the Environment in 1971 was followed by the passage of major new legislation in 1973 (the Act for the Protection of the Environment and the Land and Regional Planning Act). Environmental policy gained increasing attention from Dutch politicians, especially among political figures on the Left (Tellegen 1981). This official recognition of the environmental issue by a number of European governments stimulated demands for further environmental action.

The New Environmental Movement

The overlapping influence of multiple forces produced a dramatic surge in the environmental movement beginning in the late 1960s and continuing into

MOVEMENTS IN CONTEXT

the 1970s. One of the consequences was the establishment of organizations that represented a new ecologist consciousness born during the protest movements of the 1960s. These groups are normally identified with the predictions of NSM theory, because they generally presented a challenge both to the dominant social paradigm of advanced industrial societies and to the established political methods of Western democracies.

At the forefront of this new assault was Friends of the Earth (FoE), founded in San Francisco in 1969. David Brower, a renowned American naturalist, created the new organization after failing to convince the Sierra Club, of which he was an officer, to adopt a more activist strategy on environmental issues, especially nuclear power. Friends of the Earth wanted to address the new environmental problems of advanced industrial societies, as well as the social structures that caused these problems, with an assertive political style. Brower spread the news of his new organization on a trip through Europe the following year.[10] In 1970 several British environmentalists formed FoE-UK in London; the success of their early campaigns prompted local FoE groups to spring up independently throughout Britain.[11] That same year French environmentalists established a branch of FoE, known as Les Amis de la Terre (AdlT), with members drawn from the counterculture forces of the May Revolts. Established politicians on the Left and Right criticized AdlT for politicizing the environmental issue and for its American origins (some critics even suggested that this meant it must have CIA ties); but AdlT rapidly attracted public support. The Dutch affiliate of FoE, the Society for the Defense of the Environment (Vereniging Milieudefensie, or VMD), was established in 1972 in reaction to the conventional tactics and moderate goals of the existing Dutch environmental groups. VMD soon became nationally prominent for its actions on nuclear power, recycling, resource management, and other modern environmental issues. FoE groups gradually spread to most other European democracies and many non-Western nations. By the end of the 1970s, FoE International claimed affiliates in more than two dozen nations, including Ireland (formed in 1974), Belgium (1976), Italy (1977), and Greece (1978).

Friends of the Earth illustrates a new model of citizen action on environmental issues. FoE politicized environmentalism by discussing issues that government officials and established conservation groups often ignored, such as nuclear power, industrial pollution, and quality-of-life issues. In debating these topics, moreover, FoE groups frequently challenged the predominant social paradigm of Western industrial democracies: industrial pollution occurred because big business had thwarted control mechanisms; nuclear power represented the dominance of technocracy and large corporate interests over small-scale renewable forms of energy; waste problems arose from humans' indifference to the limits of resources. More than just environmentalists, FoE activists were social critics. Their tactics combined confrontation with authorities and events with the intent of sparking public interest. The environmental crisis called for immediate action, and FoE existed to stimulate this action.

Greenpeace constitutes another important international network of ecology groups (Hunter 1979; Brown and May 1989). Political activists formed the original Greenpeace organization in Canada in the early 1970s to protest a planned nuclear test by the United States on the Aleutian island of Amchitka. The protest united both environmentalists (green) and opponents to nuclear weapons (peace), giving the organization its name. In the next major campaign David McTaggart, using his own boat, led Greenpeace in opposing French nuclear testing in the South Pacific. McTaggart, who eventually became head of the organization, moved to Paris in 1974 to sue the French government for injuries he suffered when French naval personnel tried to expel him from the testing zone and in doing so created a bridge between Greenpeace activities in North America and Europe. Working from his Paris base and with the support of FoE chapters in Europe, he helped to establish a Greenpeace office in London and Paris in 1977. Greenpeace affiliates then quickly spread across the map of Northern Europe: Holland (1978), Belgium (1979), Luxembourg (1980), Denmark (1980), West Germany (1980), Sweden (1983), Norway, Italy (1986), Spain (1987), and Ireland (1988).

Other national groups advocating this new ecological orientation emerged during the 1960s and 1970s. In Britain, the Conservation Society, formed in 1966, advocated population control and other measures to produce a British society that could sustain itself with available resources; it also displayed its populist goals in a decentralized, antibureaucratic organizational structure (Cotgrove 1982). The ecology movement developed an especially strong organizational base in Holland (Tellegen 1981; Jamison et al. 1990). Student activists formed Action Group Straw (Aktie Strohalm) in 1970 to focus greater public attention on the social aspects of environmental problems; it now works with the Foundation for Environmental Education (Stichting Milieu-educatie, or SME), which develops educational and public information materials that stress a holistic view of nature and environmental problems. Activists established the Small Earth Society (De Kleine Aarde) in 1972, which maintains that only a change in life-styles can resolve problems such as pollution or energy shortages. The Foundation for Alternative Life (Stichting Mondiaal Alternatief) promotes the concept of a sustainable society, where resources and societal needs are brought into balance.

German environmentalists developed an extensive network of new groups, often beginning with local citizen action groups, or *Bürgerinitiativen*. In 1971, the Federal Association for Citizen Action Groups for Environmental Protection (Bundesverband Bürgerinitiativen Umweltschutz, or BBU) united several hundred of these local groups and quickly established itself as a representative of this new wave of environmental activists. Other groups, such as Robinwood, formed in 1982, mount campaigns to draw public attention to environmental problems. Research centers, such as the Öko-Institut, provide a research base for environmental action. In Denmark, an informal weekly colloquium between students and university professors evolved into NOAH, a loosely knit federation of local groups that advocate an alternative viewpoint on environmental issues (Jamison et al. 1990). In summary, the mobilization wave of the 1960s and 1970s created a new generation of European environmental groups dedicated to a more ideological and politicized view of environmental policy and willing to use more assertive tactics in support of their cause.

A Conservation Expansion

In addition to the creation of new ecology groups, the environmental surge of the 1970s stirred a parallel growth in support for traditional nature conservation organizations. During that decade the membership rolls of the RSPB virtually exploded, increasing from an average of about 10,000 members in the 1950s to 30,000 by the end of the 1960s, 100,000 in 1971, 200,000 in 1975, and more than 400,000 members by the mid-1980s. Membership in Britain's National Trust skyrocketed to nearly 1,000,000 members by the end of the 1970s. Similarly, the membership rolls of most other major European conservation groups increased dramatically during this decade (the Dutch Association for the Preservation of Natural Monuments boasted more than 200,000 members, and the Danish Association for Nature Conservation more than 100,000). The German DVB increased its membership rolls from approximately 35,000 in 1970 to over 100,000 by the end of the decade. In France, these forces led to a reorganization of conservation groups under the new structure of the French Federation of Societies for the Protection of Nature (Fédération Française des Sociétés de Protection de la Nature, FFSPN), which consolidated several dozen groups, with a combined membership of over half a million; activists founded a variety of other environmental groups during this period. Because of the regional nature of Belgian politics, separate environmental federations formed in Flanders, Wallonia, and Brussels in the early 1970s. Membership in the newly established World Wildlife Fund organizations throughout Europe also shot upward. These different groups were mobilizing citizens of various political viewpoints into specific sectors of the environmental movement.

This mobilization period also helped to institutionalize the environmental movement in Mediterranean Europe (Diani 1988, 1990a; Agnelli Foundation 1985). In 1975 Italian conservationists established the Fund for the Italian Environment (Fondo per l'Ambiente Italiano) following the model of the British National Trust. Its efforts to protect Italy's cultural heritage reinforce the activities of Italia Nostra and Pro Natura. The ecological wing of the movement is represented by the League for the Environment (Lega per l'Ambiente–Arci), the Italian branch of FoE, and the League for the Abolition of Hunting (Lega per l'Abolozione Caccia) (see also Diani and Lodi 1988). The major Greek conservation organization, the Hellenic Society for the Protection of Nature (HSPN), was joined in the 1970s by several other environmental groups and research institutes (Stevis 1990).

Retrenchment and Resurgence

In the late 1970s, following the surge in environmental action, government efforts for environmental protection began to wane. A worldwide recession shifted the attention of political elites toward economic problems, and many government officials and representatives of the established political parties saw environmental protection as a potential brake on economic recovery. Indeed, some environmentalists challenged the belief in economic growth as the driving force of social progress, a belief that is shared by traditional economic interest groups and political parties of the Left and Right. When environmentalists opposed a nuclear energy plant, for example, they often faced a united front of business leaders, union officials, and their party allies. The French government committed itself to a nuclear power program as the basis of national energy independence (Nelkin and Pollack 1981); a pragmatically oriented West German government headed by Helmut Schmidt lacked the enthusiasm of the previous administration for environmental reform and allowed environmental legislation to stagnate. In some cases, government ambiguity on environmental policy seemingly turned to enmity. Margaret Thatcher's government, elected in 1979, was openly hostile to most environmental initiatives; conservative Dutch governments in the 1980s repeated this pattern of environmental retrenchment. In 1985, agents of the French government bombed a Greenpeace ship in New Zealand, killing one crew member and illustrating the failure of government in the worst possible terms. The Greenpeace bombing was an extreme case of overreaction by those in power, but European governments and the established political parties generally became unresponsive to the environmental movement as a consequence of renewed economic problems and mounting opposition from traditional economic interest groups.

At the time, many political observers interpreted this withdrawal of support for environmental reform as a weakening of the environmental movement, but this has clearly not occurred. The mobilization wave of the preceding decade created an infrastructure of environmental groups that continued to draw attention to these issues and press for environmental action. The popular concerns that initially gave rise to these groups were not erased by economic recession; in fact, the debate over measures to stimulate economic growth often sensitized the public to the costs of economic development. Moreover, when governments were unwilling to deal quietly with environmental issues, the assertive style of environmental groups kept these issues before the public. Similarly, when the leadership of the established parties ignored the environmental issue, small minor parties catered to the environmental constituency or new green parties formed. As these Green parties won representation in their national legislatures, first in Belgium and West Germany, this provided another political forum for the environmental movement.

The efforts of established political leaders to downplay or discount the claims of environmentalists were also undercut by a continuing stream of environmental crises. In July 1976, a chemical plant in Seveso, Italy, exploded, exposing thousands to cancer-causing dioxins. The supertanker *Amoco Cadiz* broke in half off the coast of Brittany in March 1978, spewing 68 million gallons of crude oil into the sea. The spill killed thousands of seabirds and severely damaged other aquatic life, and it took several years for the environment of the Brittany coast to return to normal. The accident at Three Mile Island in 1979 created so much doubt about the use of nuclear reactors in those European nations stressing energy independence that citizens often protested against government policy (e.g., Nelkin and Pollack 1981; Hatch 1986; Rüdig 1990). One could travel to the Alps, the Black Forest, or Scandinavia in the early 1980s and see how acid rain was causing the death of the forests. In 1986, a chemical spill at a Swiss industrial firm killed millions of fish in the Rhine River and threatened local drinking water supplies from Switzerland to Holland. In addition, local development projects (such as the Rhine-Danube Canal or proposed airport expansions) or past errors (hidden toxic waste dumps) incited continuing environmental action at the local level throughout Europe. This seemingly unending stream of events in the media kept the public aware of Europe's persisting environmental problems so that environmentalism did not slip from the political agenda.

The nuclear disaster at Chernobyl in 1986 may be the event that finally changed the course of contemporary European politics on environmental issues. When the fallout spread to Western Europe, government agencies in West Germany told parents to keep their children indoors and to avoid certain foods; border guards throughout Europe checked shipments of agriculture produce with Geiger counters; even the reindeer herders in Lapland could not eat venison because of the excessive radioactivity. Scientists estimated that between a few thousand and tens of thousands of Europeans will eventually die from cancers produced by radiation from Chernobyl (Gould 1990). This nuclear accident thus created widespread public recognition that acid rain, dying forests, a decimated ozone layer, or nuclear fallout are threats comparable to the economic problems facing advanced industrial societies. Moreover, these environmental problems pose threats that cannot be avoided by individuals because of a privileged social position or geographic location. In short, Chernobyl and its by-products convinced many Europeans that the environmentalists' claims were not mere political rhetoric, and this forced political leaders to respond to public demands for environmental reform.

Over the past few years, Western governments have displayed a renewed awareness of the environmental issue. The Commission of the European Communities declared 1986–87 the European Year of the Environment, and the Single Europe Act of 1987 modified the Treaty of Rome to grant the Community new legislative authority on environmental issues. Individual national governments also became visibly active on environmental matters (Vogel 1988, 1993). The government of Helmut Kohl in Germany instituted new measures to protect environmental quality and pressed the Community to adopt policies aimed at reducing acid rain. The Greek government finally took strong measures to address Athens's extreme smog problems, passing legislation to limit automobile traffic in the center of the city. Even the Conservative government in Britain gradually responded to the pressure for environmental reform, and in late 1988 Margaret Thatcher declared herself a born-again environmentalist (repeating a conversion that George Bush had undergone a few months earlier during the American presidential elections). Green parties posted broad advances in the 1989 European parliamentary election, winning more than 10 percent of the vote in Britain, West Germany, France, and Belgium; the rainbow coalition within the new parliament includes deputies from ten nations. Western leaders toasted their newly found environmentalism by transforming the Paris Economic Summit of 1989 into an Environmental Summit (*Economist*, July 15, 1989).

Environmentalism has seemingly become a global issue in the 1990s. In addition to the environmental concerns discussed above, the problems of global warming, the ozone hole, the destruction of tropical rain forests, and other issues captured citizens' interests. The 1992 United Nations Conference on global environmental issues (known as the Rio Summit) symbolized this new concern, with more than 120 nations, as well as a full range of environmental and related interest groups, represented at the meetings (Soros 1994).

It is likely that the ebb and flow of environmentalism will continue during the 1990s, as Western democracies confront these problems and political conditions change. Moreover, because solutions will be difficult, attention to environmental issues will not be sufficient to ensure their resolution. In other words, the issues that stimulated the contemporary environmental movement will be difficult to resolve and are likely to remain potential sources of social and political conflict for a substantial period. Equally important, the successful institutionalization of the movement, in its various forms, provides the basis for the continuation of environmentalism as a force in European politics.

Two Shades of Green

The history of environmentalism has included two distinct waves of environmental mobilization. The first wave, at the turn of the century, focused public attention and organizational efforts on issues of wildlife protection and preservation of a nation's cultural and natural treasures. The second, during the 1960s and 1970s, addressed new environmental problems and quality-of-life concerns of advanced industrial societies.

In broad terms, these two eras have much in common. Both mobilizations suggest that an unusual combination of social ferment and political events might be required to create new political organizations (see, e.g., Brand 1990; Tarrow 1983; Berry 1977, chap. 2); changes in social conditions, and even more so in social climate, also contributed. In addition, institutions and individual entrepreneurs facilitated this mobilization for political action. The convergence of these factors during two discrete periods enabled the environmental movement to overcome the hurdles of institutionalization that face any prospective group and to develop an organizational infrastructure.

These two periods share similar consequences as well. Both waves were truly international in scope, representing a reformist wave that washed across Europe and North America, diffusing new issue interests and new interest groups across national borders. And both spawned groups concerned with similar collective goods: the protection of nature, preservation of fauna and flora, and improvement of the quality of life. Environmental groups of all colors share a common bond when compared to economic interest groups (such as unions and business associations) or other public interest lobbies. The political interests of different environmental groups routinely overlap, even if they have separate starting points. Bird protection groups, for example, now realize that the destruction of habitats by industrial pollution should be a concern of their members, just as modern ecological groups structure their campaigns to tap widespread public support for nature conservation issues. In exploring the political activities of the environmental movement, we shall see that formal and informal patterns of cooperation interrelate a broad range of groups. Available evidence also suggests a substantial overlap in their membership and funding bases. At one level, it does make sense to talk of a single environmental movement, especially if we want to be inclusive in studying the sources and forms of environmental action.

At the same time, the two waves of environmental mobilization are not two examples of an identical phenomenon, separated only by several decades. Each produced a different orientation toward what environmentalism entails and how it relates to the prevailing social norms of Western industrial democracies. In later chapters we shall explore how the organizations spawned during these two mobilizations differ in their organizational structure, issue agendas, political allies, and choice of tactics. The crucial starting point for these other differences, we will argue, lies in the differing ideology or collective identity of the environmental groups formed during these two mobilization waves.[12]

The first mobilization emphasized a shade of environmentalism that we shall call a *conservation orientation*. At this stage, environmentalism meant a concern with the perpetuation of species, the protection of habitats, and the preservation of a nation's heritage as represented by its cultural monuments and environment. As a philatelist might save a rare stamp for his or her collection, conservationists wanted to save fauna, flora, and cultural sites for themselves and others to enjoy. The various bird societies, wildlife protection associations, and cultural preservation groups formed at the turn of the century generally reflect a conservation orientation (see classification in chapter 4). The conservation movement was supplemented by new groups that formed during the 1960s and 1970s—such as WWF, DNR, and FFSPN— and broadened public support for conservation issues.

The second mobilization wave created a new *ecological orientation* that differs from conservationism in its political concerns and fundamental political ideology. New organizations formed in the 1970s and 1980s, such as Friends of the Earth and Greenpeace, focus on the need for fundamental social change to address issues such as nuclear power, industrial pollution, acid rain, silent carcinogens, and the other new environmental problems of advanced industrial societies. The contrast between conservationists and ecologists, however, involves more than just issue interests.

Conservationists generally accept the existing sociopolitical order and the norms of this social system. They are pursuing consensual social goals within the existing socioeconomic structure; an organization need only bring attention to an unrecognized issue that society already supports. Most people, for instance, favor preservation of sites of natural beauty and the general protection of wildlife. In a similar classification of environmental groups, Philip Lowe and Jane Goyder refer to conservationists as "emphasis groups . . . whose aims do not conflict in any clear-cut way with widely held social goals or values but which are motivated by a belief in the importance of certain values and the need for vigilance on their behalf" (1983, 35). Thus, conservationists do not question the dominant values of society; rather, they seek to preserve what they believe is a threatened part of that social system. Few would question the labeling of bird protection organizations as conservation groups, even if some members and activists in these groups do criticize the prevailing social order for the destruction of habitats and feeding grounds for birds. Similarly, when an organization labels itself as a "Royal Society," there can be little doubt about its acceptance of the prevailing social order.

Ecologists, in contrast, advocate a basic change in societal and political relations as a prerequisite for addressing environmental problems.[13] The most important ideological feature of an ecology group is its rejection of the values of the prevailing social order. Ecologists supposedly are attempting to develop a new societal model—what is called a New Environmental Paradigm (Dunlap and van Liere 1978; Milbrath 1984, chap. 2; Cotgrove 1982, chap. 2). This ecological orientation represents an eclectic humanist philosophy rather than a rigid ideological framework.[14] Moreover, these values overlap with postmaterial orientations (Inglehart 1990; Rohrschneider 1990), linking environmentalism to other NSMs and to other political changes transforming advanced industrial democracies.

Instead of stressing (or even accepting) the accumulation of wealth and unrestrained economic growth as societal goals, ecologists advocate an economy in harmony with nature and the personal needs of its citizens. The writings of E. F. Schumacher (1973) often provide a beginning reference point. Ecologists roundly criticize the materialism and excess consumerism of modern societies as causes of resource shortages and sources of pollution.

Ecologists advocate a sustainable society, in which consumption patterns are balanced against the world's natural productive capacity (Brown 1981; Irvine and Ponton 1988; Milbrath 1989). These sentiments imply a restructuring of the economic system of Western industrial societies, substituting the domination of the environment with a nature-friendly economic order.

The alternative social paradigm of ecologists also emphasizes the value of the individual and the need for a more humane social order. Modern societies are highly structured and bureaucratized (a *Gesellschaft* model), but ecologists advocate a pattern of social relations that harks back to small-scale communal life (a *Gemeinschaft* model). The status differences in the dominant social paradigm are replaced by a sense of egalitarianism and the importance of individuals. This pattern of social relations allows a sense of collective identity to coexist with a tolerance of individual life-style choices. Claus Offe (1990) suggests that NSMs, such as the environmental movement, are based upon the values of personal integrity, recognition, and respect.

A stress on participation and self-direction is another trait identified with ecology groups.[15] Drawing off the traditions of Pyotr Kropotkin and Henry David Thoreau, the new environmental paradigm contends that personal fulfillment and self-expression should be maximized. Individuals must control society, rather than have society control the individual. Thus, ecologists criticize nuclear power not just for its environmental consequences but also because it signifies the domination of industry and technology over the individual. This orientation leads ecologists to advocate changes in the political structure of Western democracies, leading to a more open political system, a greater opportunity for citizen input, and a more consensual style of decision making. In contrast, conservation groups often began as small, elitist associations. Even when these groups become mass organizations, they lack the emphasis on participatory politics often found in ecological groups.

In short, ecologists should see current environmental problems not as an aberration of the socioeconomic order but as a direct consequence of this social system. Instead of trying merely to preserve parts of the natural world, as do the conservationists, ecologists are more likely to question the economic and social structures that are seen as the sources of contemporary environmental problems. Some ecologists express these sentiments in radical terms. Rudolf Bahro, an activist in Germany's Green Party, often intermixes radical rhetoric with his alternative viewpoint: "In the richest, industrially overdeveloped countries of the West a fundamental opposition is growing. . . . It is reacting to the now clearly and markedly self-destructive, outwardly murderous and inwardly suicidal character of our industrial civilization, and to its institutional system which is geared to continuing in the same old way" (Bahro 1984, 1).

Other ecologists stress a humanist sensitivity. Robert Hunter, one of the original founders of Greenpeace, describes his orientations: "We need to move neither further to the Left nor further to the Right—rather, we must seriously begin to inquire into the rights of rabbits and turnips, the rights of soil and swamp, the rights of the atmosphere, and ultimately, the rights of the planet. For these in the end are the containers of our entire future evolution, and everything rests upon whether or not we come to terms with the politics of earth and sky, evolution and transformation, God and nature" (Hunter 1979, x). In overall terms, the cultural criticism of ecologists is much more fundamental than the social criticism of the conservation movement. In later chapters we shall argue that adherence to a new environmental paradigm (or to the values of the dominant social order) is a key factor affecting the behavior of environmental groups—and thus the primary criteria to use in defining a group as holding conservation and ecological orientations.

Our framework of Ideologically Structured Action implies that the challenge to consensual social goals or dominant sociopolitical actors by ecology groups will affect their potential resource base and their relationship with other political actors. This, in turn, will influence the alliance patterns, political opportunities, political tactics, and other aspects of the organizational behavior of ecology groups. For instance, established economic and political interests may be more accepting of conservation groups as participants in the policy process because their values and methods do not challenge the existing order; lobbying and negotiation thus might be effective means of influence for these groups. In contrast, as critics of the socioeconomic system, ecology groups might find that such conventional methods must be combined with confrontational tactics.

This distinction between old and new elements of the environmental movement undoubtedly will blur when the theoretical ideal is tested against reality. Elements of fundamental social criticism existed in the environmental movement of the 1800s, just as the mobilization wave of the 1970s produced new members of the conservation movement—though in both instances the dominant forces were in the other direction. Furthermore, most environmental groups combine some mix of conservation and ecological orientations; even the various national affiliates of such international groups as WWF, Greenpeace, and FoE vary in their adherence to a common orientation. Pure ideal types exist only in the minds of social scientists. Still, these ideal types represent different tendencies that have been widely observed within the environmental movement.[16]

As we argued in chapter 1, the identity or ideology of an organization should influence both its organizational and political behavior. The distinction between conservation and ecological orientations provides a means to test this theory. These two environmental orientations represent quite different perspectives on how individuals should interact, how the political process should function, and how society should be structured. Ecology groups are regularly cited as the "new" social movements of NSM theory. In addition, our focus on environmental groups enables us to exclude other aspects of interest group politics that might distort the impact of ideology. We do not have to contend with the potential differences between public-interest and private-interest groups, for example, because all types of environmental groups share an interest in collective goods. Similarly, most mass-membership environmental groups attempt to mobilize roughly similar resources and members for their organization. Our comparisons of conservation and ecologist tendencies within the environmental movement should therefore yield a fairly pure measure of the impact of ideology on organizational behavior.

Chapter Three

EUROPEANS AND

ENVIRONMENTALISM

The historical development of the green movement provides one context for understanding contemporary patterns of political action, but another important context involves popular sentiments toward the movement and its goals. As a social movement, the growth and success of environmental groups is inevitably linked to the citizenry's attitudes toward the movement. A supportive public provides a base from which an SMO might mobilize resources and members. Social movement scholars stress that the support of a large external constituency is an extremely valuable resource in any movement (e.g., Tilly 1978). Perhaps even more important, a supportive public provides vital legitimacy for a social movement that claims to represent the public's interests. As we shall note in chapter 4, an interest group's image can be one of its most precious resources.

Reports in the popular press suggest a greening of European politics over the past decade. Green parties made dramatic progress during the 1980s, as did environmental interest groups. The environmental conversions of political figures such as George Bush and Margaret Thatcher similarly attest to the apparent popularity of the movement. (Yet the failure of these two politicians to make meaningful conversions may have contributed to their downfall.) This evidence alone suggests broad popular support for environmentalism.

There are, however, many individuals who question the depth of the public's commitment to environmental reform. Skeptics sometimes describe environmentalism as a sunshine issue that attracts public support only when more pressing (economic) issues are in abeyance. From this perspective, citizen interest in environmental issues and support for environmental reform are not deeply rooted beliefs but play a secondary role to more important economic and social concerns. Anthony Downs (1972) similarly suggested that public interest in environmentalism would, like other social problems, be a transitory phenomenon; initial public concern would wane once the issue lost its novelty and the costs associated with environmental protection measures became apparent. Other critics claim that the environmental movement is so extreme that it attracts only a narrow citizen base (Tucker 1982a, 1982b). Indeed, the disdain with which some politicians treat environmentalists is evidence of their willingness to alienate supporters of the movement. Even in a more benign form, the media often present environmentalists as a marginalized political movement of romantics and unreformed hippies.[1]

This chapter addresses these questions by mapping the attitudes of Europeans toward environmental issues and environmental groups. We begin by determining the breadth and depth of public support for environmental reform. A key part of this inquiry asks whether environmentalism is merely a passing fad or a deeply rooted commitment on the part of the public. The second section of the chapter focuses on public support for environmental groups; in it we contrast popular images of nature conservation groups with those of ecology organizations. Finally, we describe the correlates of support for environmental groups to determine the social and political roots of the movement.

Public Attitudes toward the Environment

A distinguishing feature of environmentalism is the mass base of the movement. Environmentalism moved onto the agenda of democratic polities because individual citizens took an interest in these issues, small groups formed to address local environmental problems, and citizen interest groups emerged as advocates of these new concerns.

This section traces the evolution of European public opinion toward environmental protection from the 1970s into the 1990s. Fortunately, the Commission of the European Communities has funded a long series of public opinion surveys. These Eurobarometer surveys are conducted semiannually among the populations of each EC member state.[2] In addition to general questions on European issues, they periodically include extensive items on environmental themes (Commission of the European Communities 1982, 1986). In fact, the Directorate General for the Environment commissioned a special survey on environmental issues a few months after our group survey, thereby adding to our research base.

The Salience of Environmental Protection

The most basic measure of environmentalism is the public's interest in the associated problems. Do most individuals see environmental protection as an esoteric issue removed from the everyday concerns or as a salient political problem?

Although conservation groups have been active for nearly a century, until very recently most politicians did not consider environmentalism a political issue of mass appeal. The protection of nature was an avocation of the socially elite rather than an issue of general public interest. The average citizen seemed more concerned about his or her economic well-being, social security, and other sustenance and security issues. In this context, environmentalism apparently existed at the fringe of politics. As recently as 1957 the German Social Democrats explored an environmental issue as a campaign theme ("Blue Skies over the Ruhr") and found public interest lacking.

During the past two decades, however, green issues have dramatically entered the political agenda of most advanced industrial democracies. At first, environmental crises, such as the sinking of an oil tanker or massive bird kills, captured the public's attention and provided new fuel for the movement. But now, a large proportion of the public appears interested in environmental matters as more than just crisis management. Discussions of environmental quality, energy options, protection of nature, and similar topics attract extensive interest.

The Eurobarometer surveys began asking about citizen interest in environmental protection in 1973. At that time, a large majority in each EC nation (except Ireland) already felt that environmental protection was an important or very important issue (Dalton 1984). In 1976 a new question asked respondents to rate the importance of environmental protection as a political issue (table 3.1).[3] Although this survey followed the OPEC-induced recessions of the early 1970s, when economic issues presumably became more prominent, interest in environmental protection was nonetheless widespread among Europeans. Except for Britain and Ireland, two-thirds of the public in the other EC member states rated the environment as a "very important" issue in 1976. The public's interest in environmental issues also shows in comparison to other issues. Among the eleven other issues listed in the 1976 survey (data not shown), Europeans ranked only unemployment and inflation as more important than the environment. Most ranked a host of other long-standing issues—such as housing, education, defense, consumer protection, and regional policy—as less important than the environment. In

TABLE 3.1

Public Interest in Protecting Nature and Fighting Pollution, 1976–89

Nation	1976	1978	1983	1987	1989
France	71%	61%	53%	56%	68%
Belgium	65	54	46	57	76
Netherlands	67	70	53	61	83
West Germany	64	55	64	69	83
Italy	67	62	58	68	85
Luxembourg	75	42	66	73	78
Denmark	74	64	79	85	89
Ireland	49	47	37	49	71
Britain	48	48	48	53	75
Greece	—	—	68	67	71
European Community	63	57	56	62	78

Source: Eurobarometer 6 (October 1976), 10 (October 1978), 20 (October 1983), 28 (October 1987), and 31a (June 1989).
Note: Figures are the percentage saying that the issue is "very important," the highest of four responses to the question.

short, by the mid-1970s most Europeans were already acknowledging the importance of environmental protection.

Public attention to environmental matters remained widespread in Europe over the next decade, despite the revival of economic worries in the late 1970s and early 1980s. Opinions fluctuate in specific nations, but most Europeans rated environmental protection as a very important problem in the next three surveys. Then, in the late 1980s attention rose dramatically. In 1989, a full 78 percent of Europeans saw the environment as a very important problem (table 3.1), and further data (using differently worded questions) show that the rise in Europeans' environmental interest continued into the early 1990s (*Eurobarometer 37*, 1992, 66–67). Even in Britain and Ireland, where environmental interest initially was modest, more than two-thirds of the public saw the environment as a pressing problem by the end of the 1980s.[4]

Other data confirm this trend of increasing popular interest in environmentalism. For instance, the German election surveys conducted by the research firm Forschungsgruppe Wahlen found that the proportion of voters

MOVEMENTS IN CONTEXT

saying that the environmental protection issue is very important increased from 43 percent in 1972 and 52 percent in 1980 to 69 percent in 1987 (Conradt and Dalton 1988, 14). Even in the midst of German unification, environmental protection was the third most important issue to the public during the 1990 elections (Forschungsgruppe Wahlen 1990). Longitudinal data from the Dutch Continuous Survey similarly described a consistently high concern with environmental problems from the late 1970s to the 1980s. Data from the MORI poll in Britain also shows that public attention to environmental issues rose dramatically in the late 1980s (*Economist*, Mar. 3, 1990, 49).

These findings appear to refute the claim that popular interest in environmentalism is merely a sunshine issue. Rather, it is a priority concern for most Europeans, and this sentiment has endured through major recessions in the 1980s. If anything, environmental interest grew during the 1980s and continued its rise at the beginning of the 1990s (see, e.g., membership statistics in chap. 4). Moreover, Jürgen Hofrichter and Karlheinz Reif (1990) show that even when ranked against other pressing political goals, environmental protection routinely appears to be a top priority with each populace. Only unemployment evoked greater concern from Europeans during the late 1980s, and in most nations people ranked the environment among the two or three most important problems. Environmental activism and public interest have expanded the boundary of politics to include green issues on the political agenda, and popular concern for these issues appears durable.

Support for Environmental Protection

As skeptics point out, it is relatively easy for individuals to say that environmental protection is an important problem when a interviewer comes knocking at the door, but it is less certain that this interest can be translated into support for meaningful environmental reform. Moreover, we have stressed that there are many shades of green, and concern for the environment may mean protecting birds to one individual, cleaning up a local environmental problem to another, and restructuring life-styles to encourage a sustainable society to a third. An expressed interest in green topics does not therefore indicate the depth of one's environmental commitment.

We do not attempt to describe the full range and structure of public beliefs about the environment (see, e.g., Milbrath 1984; Cotgrove 1982). Instead, we assess the depth of popular support for environmental protection as a general policy goal. When environmental protection is considered by itself, Western European publics are nearly unanimous is expressing their

TABLE 3.2

Support for Environmental Reform Issues

Nation	Favor Environment over Prices	Favor Environment over Growth	Favor Protecting Environment over Economy
France	63%	58%	56%
Belgium	50	50	35
Netherlands	72	56	45
West Germany	54	64	50
Italy	66	67	55
Luxembourg	69	64	65
Denmark	74	75	55
Ireland	34	29	40
Britain	57	50	48
Greece	67	56	47
European Community	60	59	52

Source: First column is taken from Eurobarometer 18 (October 1982); the next two columns are from Eurobarometer 25 (May 1986).
Note: Figures are the percentage of the populace agreeing with the statement. Missing data included in the calculation of percentages.

environmental concern. For instance, several Eurobarometer surveys have asked whether Europeans agree that "stronger measures should be taken to protect the environment against pollution"; more than 95 percent agreed (Dalton 1984)! The skeptics note, however, that such questions are unlikely to evoke negative opinions. Who, it is asked, would oppose clean air, clean water, and a better environment? Difficulties arise when protecting the environment comes into conflict with other valued goals. Will the public still support strict air quality standards if these regulations raise their utility bills or require expensive lead-free gasoline? Such trade-offs provide a more stringent test of the public's real commitment to environmental protection.

A 1982 Eurobarometer survey asked respondents to balance their environmental beliefs against the potential economic costs of environmental protection (table 3.2).[5] Although the survey follows on the heels of an EC-wide recession in the early 1980s, most Europeans attach higher priority to environmental protection, even when these programs are linked to potential inflation or a slowdown in economic growth. In France, for example, 63 percent of the public favor protection of the environment even if it causes companies to raise their prices; 58 percent feel that environmental protection should take priority even if it risks economic growth. These French percentages are fairly close to the European average. Only in Ireland does less than a majority favor the environmental options. When the question linking environmental protection to environmental growth was repeated in a subset of nations in 1992, environmental support had become even more widespread.[6]

Eurobarometer 26, conducted a few months after our environmental group study, asked a variant of this trade-off question: Should economic growth take priority over environmental protection?[7] Once again a majority of Europeans endorse the environmental option, this time implying both a preference for environmental quality and an assumption that environmental quality contributes to economic growth (column 3 of table 3.2).

Most Europeans certainly prefer both economic growth and a clean environment. If forced to choose between these goals, however, a strikingly large number state a preference for environmental quality; this finding has now been widely replicated in other surveys of Western publics (Milbrath 1984, 27; Dalton 1988, chap. 6; Gallup Institute 1992). Economic costs obviously could be increased to the point where the balance shifts away from environmental protection. Still, the expressed support for the environment in the face of potential economic losses indicates the depth of the public's environmental beliefs. Moreover, these patterns probably differ from the distribution of opinions we would have uncovered a generation earlier.

A final measure of the behavioral depth of European's concern about the environment comes from a question asking what actions individuals had taken to protect the environment (table 3.3). Most have already taken several steps, such as preventing litter (77 percent), controlling noise (50 percent), and conserving water (44 percent). Many others (45 percent) claim to be recycling as many household products (glass, paper, oil, and the like) as possible. In nations with active recycling programs, such as Luxembourg, Denmark, and Germany, more than half of the public claims to participate in recycling activities. Equally impressive are the numbers who have taken part in more demanding activities. More than one-tenth of Europeans have contributed money to protect the environment; significant numbers say they have voluntarily taken steps to reduce the emissions of their cars, have participated in a local effort to clean up the environment, or have been involved with an environmental association. Admittedly, only a minority say they have engaged in these additional activities, but this is a relatively high level of involvement compared to many other voluntary political actions.[8]

TABLE 3.3

Individual Activities Undertaken to Protect the Environment

Nation	Avoid Littering	Save Water	Limit Noise	Convert Auto Exhaust	Donate Money	Recycle	Join Local Action	Protest	Work with Assoc.	Avg. Acts
France	84%	47%	61%	5%	6%	45%	5%	7%	6%	2.7
Belgium	69	54	48	4	13	44	5	5	6	2.5
Netherlands	74	48	55	5	39	57	5	6	14	3.0
West Germany	64	42	36	15	13	63	12	4	2	2.5
Italy	86	51	57	2	7	32	7	9	8	2.6
Luxembourg	85	57	.64	3	25	70	17	9	14	3.4
Denmark	71	21	32	8	21	40	6	7	16	2.2
Ireland	75	35	33	3	9	12	5	4	4	1.8
Britain	86	48	52	3	16	33	5	4	7	2.5
Greece	70	38	46	4	4	2	6	8	3	1.8
European Community	77	44	50	7	12	45	6	6	7	—

Source: Eurobarometer 25 (May 1986).

Note: Figures are the percentage of the populace saying they had already done the activity. Missing data included in the calculation of percentages. The last column is the average number of acts performed by citizens in that nation.

Other evidence points to the public's willingness to translate their expressed concern for the environment into changes in behavior. John McCormick (1991, chap. 6) describes the rise in green consumerism in Europe. Green consumer guides have become publishing successes in Europe and the United States, and sales of "environmentally safe" or "environmentally friendly" products have noticeably increased (even if the meaning of these terms remains ambiguous to consumers). Businesses are finding it profitable to boast of their green credentials. McCormick further observes: "A MORI poll in June 1989 revealed that more than 18 million people, nearly half the adult population of Britain, considered themselves environmentally conscious shoppers. . . . It further found that 42 percent of people questioned met the MORI definition of a 'green consumer': someone who had made at least one purchase in the previous twelve months where one product was selected over another because of its environmentally conscious packaging, formulation, or advertising" (McCormick 1991, 108).

In the European nations surveyed by the Gallup *Health of the Planet Study* (1992), large majorities claim that they avoid using certain products that harm the environment: Germany, 83 percent; Denmark, 65 percent; Great Britain, 75 percent; the Netherlands, 68 percent; and Ireland, 63 percent. Other studies find that about one-fifth of the German and British public say they are willing to pay more taxes to protect the environment (Kessel and Tischler 1982, 96). Similarly, a majority of the British and German public say the government should spend more on environmental protection (Eurobarometer 21). The electoral success of green parties is another hard indicator of green action, although it underestimates the potential electoral impact of environmentalism. Survey evidence from the Eurobarometers indicates that nearly half of the European public would be willing to vote for a green party (Inglehart 1990, 266). Finally, German public opinion surveys have found that the public would accept speed limits on the autobahn if this were necessary to limit pollution; for German automobile owners, there could be no greater sacrifice.

In summary, the data presented here describe a European public that is broadly interested in environmental protection and expresses a willingness to take significant actions to ensure and improve the quality of the environment. Although we should be skeptical that all who express support for environmental reform will agree when real choices are being made or real costs are incurred, these statements of support are still meaningful indications of how Europeans view environmentalism. There are some national variations in environmental support—it is somewhat higher in Denmark and Luxembourg and lower in Britain and Ireland—but by the end of the 1980s the overall patterns speak of broad support that crosses national boundaries. Moreover, rather than being an issue that fades when economic problems revive, public support for environmental reform has endured over the past two decades.

Opinions of Environmental Groups

Europeans' support of environmental reform should lead them to evaluate positively the prime advocates for reform: environmental groups. This relationship is not assured, however. Sympathizers of environmentalism emphasize the broad popular appeal of green issues, as we have seen, and attribute this same appeal to green groups. Yet critics of the movement stress the extremism of some environmentalists and suggest a marginalized status for environmental groups because of their unconventional image.

In addition, there are many shades of green, and the nature of a particular environmental group should affect its popular appeal. Who, for example, could be critical of the Royal Society for the Protection of Birds or the Civic Trust in Britain? Yet one might expect the average Briton to have ambivalent feelings toward groups such as Friends of the Earth and Greenpeace, both of which challenge the established political order in their goals and methods. It is therefore important that we distinguish among the various components of the green movement.

Paralleling our theoretical distinction between conservation groups and ecology groups in the preceding chapters, the Eurobarometer surveys asked people their opinions of both types of groups. Because of the diversity of groups and the difficulty of distinguishing among them in a general population survey, the questions mentioned "nature protection groups" or "the ecology movement" rather than specific organizations (see also Rohrschneider 1989).[9] Europeans are almost unanimously positive toward nature protection groups, with nearly nine-tenths expressing approval (table 3.4). Even the small percentages that are unaccounted for in these statistics are more likely to have no opinion than to disapprove of conservation groups. Regarding traditional nature protection groups, virtually all Europeans are greens.

The surveys also asked respondents about their opinions of the ecology movement. Popular support exists for ecology groups, but with less approval than for nature conservation groups. On the one hand, this modest support is in keeping with the challenging nature of ecology groups; public attitudes toward antinuclear power groups are similarly mixed (Inglehart 1990, chap. 11; Rohrschneider 1990). On the other hand, more than two-thirds of Europeans approved of the ecology movement in 1982, and approval ratings rose even higher toward the end of the decade (79 percent in the five-nation subset in 1989).

The Eurobarometer surveys also asked whether respondents were members of either type of group, and if not, whether they would be willing to become members.[10] In 1986, for instance, 35 percent of the European public expressed a willingness to join a nature conservation group, and 19 percent said they would join an ecology group. The potential membership of green groups remains fairly stable across the 1980s, rising by 4 percent for nature protection groups and 6 percent for ecology groups (for these data see Rohrschneider 1989; Fuchs and Rucht 1990).[11] Still, even these modest percentages translate into a potential corps of millions of environmentalists

TABLE 3.4

Approval of Nature Protection and Ecology Groups, 1982–89

Nation	Nature Protection Groups				Ecology Groups			
	1982	1984	1986	1989	1982	1984	1986	1989
France	92%	91%	96%	92%	70%	76%	73%	83%
Belgium	78	—	85	—	67	—	73	—
Netherlands	90	90	91	96	83	80	87	89
West Germany	73	81	86	87	41	44	48	58
Italy	93	93	93	96	86	86	86	89
Luxembourg	94	—	95	—	65	—	84	—
Denmark	81	—	85	—	40	—	47	—
Ireland	84	—	74	—	51	—	56	—
Britain	94	81	89	92	60	66	65	74
Greece	88	—	81	—	72	—	67	—
European Community	90	—	90	—	63	—	69	—
5-nation average	88	87	91	93	68	70	72	79

Source: Eurobarometer 17 (April 1982), 21 (April 1984), 25 (May 1986), 31a (May 1989).
Note: Figures represent percentage of the populace saying they "approve" or "strongly approve" of nature protection groups and ecology groups; dashes indicate that the question was not asked in that nation. No opinion and missing data responses are included in the calculation of percentages.

and yield an impressive display of support when compared to other voluntary political activities.

To summarize citizen support for environmental groups in our set of nations, we combined the psychological (approval) and behavioral (membership) dimensions of group evaluations into a single scale. The highest level of environmental support is actual membership in an environmental group, followed by willingness to join, two levels of approval, indifference and, finally, two levels of disapproval (table 3.5).

The overall patterns for nature protection groups and ecology groups repeat what we have already stated: support is greater for nature protection groups. In addition, however, these data display important cross-national differences in support for environmental groups. The Danish public, for instance, is among the most active in nature protection groups; this reflects the success of groups such as the Danish Association for Nature Conserva-

TABLE 3.5

Public Support for Nature Protection and Ecology Groups, 1986

	FRN	BEL	NET	GER	ITA	LUX	DEN	IRE	BRI	GRC	EUR
Nature protection groups											
Member	1	3	9	2	2	15	15	1	3	1	3
Potential member	20	7	36	59	24	29	17	28	30	36	33
Strong approval	45	40	32	10	38	33	36	25	24	35	30
Approval	30	39	17	17	30	18	22	28	37	10	27
Indifference	1	7	2	9	3	2	6	14	3	15	5
Disapproval	1	3	3	3	2	2	3	3	2	1	2
Strong disapproval	1	2	2	1	1	1	2	2	1	2	1
Total	99%	101%	101%	101%	100%	100%	101%	101%	100%	100%	101%
Ecology groups											
Member	<1	1	3	1	1	5	<1	1	1	<1	1
Potential member	10	5	28	25	17	19	6	19	17	30	18
Strong approval	24	35	31	8	33	31	33	24	24	31	23
Approval	40	38	27	18	37	31	23	28	37	11	31
Indifference	4	8	2	10	5	5	21	22	11	24	8
Disapproval	17	8	5	23	4	6	9	4	7	2	12
Strong disapproval	5	5	3	16	3	3	8	2	3	2	6
Total	100%	100%	99%	101%	100%	100%	100%	100%	100%	100%	99%
(*N*)	1,003	1,007	1,001	987	1,102	299	1,043	1,002	1,055	1,000	10,148

Source: Eurobarometer 25 (May 1986).

tion, which claims a member in one out of ten Danish households, as well as other conservation groups, such as WWF. Dutch membership in nature protection groups is also significantly higher than the European average, undoubtedly based on the large mass membership of Natuurmonumenten, Heemschut, WWF, and other Dutch conservation groups. Outright opposition to nature protection groups is very limited; no more than 5 percent of the public in any of the countries declare disapproval. More common are expressions of indifference toward these groups, especially among the Irish (14 percent) and Greeks (15 percent). In both nations we would interpret public indifference as an indication of the low visibility of environmental groups.

The data also show that membership and potential membership in ecology groups lag behind citizen support for nature conservation groups. In overall terms, the strongest base of the ecology movement is probably the Netherlands. Nearly one-third of the Dutch public are willing to join an ecology group, and an additional 58 percent express approval of ecology groups. Thus for both nature conservation groups and ecology groups, the Netherlands appears a fertile ground for the environmental movement.

In contrast, German opinions toward the ecology movement are highly polarized. Potential membership in ecology groups ranks above the European average, but so too does outright disapproval of these groups. A full 39 percent of Germans disapprove of ecology groups, by far the largest negative percentage in any nation. This polarization of opinions might reflect the more active and politicized nature of ecology groups in the Federal Republic and the reference to the Green Party as an example of the ecology movement in the interview protocol. To a lesser extent, French opinions toward ecology groups are also polarized; whereas 35 percent of the public at least strongly approve of ecology groups, another 22 percent express disapproval.

There are three other nations where support for ecology groups appears to lag significantly behind the European average: Belgium, Denmark, and Ireland. Each case can be explained in different terms. In Belgium, few individuals express willingness to join an ecology group (6 percent), probably because of the politicized and unconventional style of the Belgian ecologists (Kitschelt and Hellemans 1990). However, the Belgian public has a favorable opinion of ecologists (73 percent express approval). The Danish and Irish statistics are surprising in other ways. Danish ecology groups have been among the most successful in Europe in attracting members and gaining political influence; Ireland, in contrast, has only a very small, and not very visible, ecology movement. In both nations, however, large proportions of the public say they do not have an opinion about the ecology movement, and this, rather than negativism, accounts for the lower percentages of approval.

The evidence we have presented points to a European public that is broadly supportive of the environmental movement. This is not surprising for nature protection groups, because their goals, tactics, and political position are largely accepted by European societies. It is more striking that this support extends to ecology groups, which are more likely to challenge the goals and conventional style of the political establishment. The breadth of support for green groups points to an inherent strength of the environmental movement—environmentalists can cite such polls as evidence of the validity of their claims for environmental reform. Such support also implies that attempts to mobilize members, solicit contributions, and otherwise assemble resources for the movement will find a receptive public. There is, in other words, a broad basis for environmental action in these European societies.

The Patterns of Support for Environmental Groups

The descriptive information in the previous section helps to define the general political context for environmental action. These results point to a sympathetic public that is willing to endorse the green movement and its goals— at least as abstractly measured in public opinion surveys.

We can gain an even better understanding of the immediate and long-range significance of environmental attitudes by determining the distribution of these opinions within the European public. Who are the environmentalists, and why do they support environmental action?

We shall focus on two general factors in examining the political bases of environmentalism. First, many analysts stress that support for the green movement is conditioned by an individual's social location, that is, factors such as class and education. One of the most common explanations of environmentalism traces the roots of the movement to an expanding group of university graduates and the growth of the new middle class (Cotgrove and Duff 1980).[12] There are different theoretical explanations of this social pattern. From one perspective, the affluent and well-educated middle class is poorly integrated into the traditional bourgeois or proletarian framework of industrial societies, in both social and psychological terms. This upper stratum becomes a promising source for mobilization by the environmental movement and other NSMs because it is less concerned with the economic matters that underlie traditional class-based political issues (Bürklin 1987b; Rohrschneider 1990). In addition, these individuals are more likely to hold political values that would stimulate quality-of-life concerns and activism in reformist movements such as the green cause (Inglehart 1977, 1990). Indeed, the leadership of the nature conservation movement is often drawn from the social elite in Europe; the origins of the ecology movement frequently can be traced to protest groups among university students.

Other researchers maintain that environmentalism reflects a political syndrome of middle-class radicalism. These scholars argue that the green movement derives from a political critique of the goals and institutions of industrial societies by reformist or radical elements of the middle class. Sometimes these sentiments are linked to a political ideology of leftist (or

New Left) opposition; such radical critiques are often pronounced in the writings of prominent environmentalists (e.g., Bahro 1984; Capra and Spretnak 1984; Kelly 1984; Pepper 1984; Manes 1990). Other times these views are tied to the thwarted ambitions of a growing number of university-educated youth who feel frustrated by the lack of opportunities for future advancement (Bürklin 1987b; Alber 1989). Claus Offe (1984, 1985) has similarly linked NSMs to marginalized social groups, in which he includes students. Even critics of the green movement stress its middle-class radicalism, pointing out the contradiction that those who have benefited by the material advantages of advanced industrial societies now seem willing to restrict those benefits (for others) in the name of environmental protection (Tucker 1982a). A study of the environmental movement by representatives of the nuclear power industry refers to environmental activists as the "penthouse proletariat" or "goose-liver proletariat" whose actions reflect their willingness to protest the political order while occupying a position of privilege (Foratom 1979).

The empirical evidence on social differences in environmental support is mixed. Cotgrove and Duff (1980) found sharp class differences in support for environmentalism in Britain, and even distinct differences between market and nonmarket sectors of the middle class. Robert Rohrschneider (1990), on the other hand, found only weak class differences in support for environmental groups across the European public. Education appears to be a stronger predictor of support for the green movement, though its effects appear most noticeable among the young (Bürklin 1984; Rohrschneider 1990). Our analyses of the social base of environmentalism therefore will examine class, education, and generation as potential determinants of support for the movement.

A second general explanation of public support for environmentalism stresses psychological factors. The most far-reaching theory focuses on the changing value priorities of Western publics. Ronald Inglehart has persuasively argued that the social transformation of advanced industrial societies has led to parallel changes in public values (Inglehart 1977, 1990). For several decades following World War II, Western democracies experienced an unprecedented growth in individual and societal affluence, coupled with the expansion of the social welfare state. Having grown up in an era when traditional economic and security goals seemed relatively assured, younger generations are shifting their attention toward noneconomic, or postmaterial, goals. Compared to their elders, many young people place a higher priority on self-expression, personal freedom, social equality, self-fulfillment, and maintaining the quality of life. Inglehart has assembled an array of public opinion evidence, impressive in its cross-national and cross-temporal breadth, that documents these changing value priorities. Postmaterial values should encourage support for environmentalism, because these values emphasize qualitative social goals over economic security and growth. In addition, postmaterial values are conducive to the participatory and direct action orientations that are another hallmark of the environmental movement (Inglehart 1990, chap. 10). Therefore, the postmaterial thesis views environmentalism as an evolutionary consequence of the social transformation of advanced industrial democracies and thus as an enduring feature of our political future.

Prior empirical research has documented a strong relationship between postmaterial values and environmental attitudes. Cotgrove and Duff (1981) found that postmaterialists were often adherents of the cluster of beliefs described as the New Environmental Paradigm and were disproportionately represented among the membership of British ecology groups. Postmaterialism is a potent predictor of support for green parties (Müller-Rommel 1990; Rohrschneider 1993b). More directly related to our interests, both Inglehart (1990, chap. 11) and Rohrschneider (1990) find that postmaterial values emerge from multivariate statistical models as a major predictor of support for environmental interest groups. The Eurobarometer series contains an index of value priorities that enables us to assess the link between values and support for nature protection and ecology groups across nations.[13]

In addition to this process of postmaterial value change, citizens' broader political orientations also shape their opinions toward the green movement. Left/Right orientations, for instance, are at the core of many political schema (Conover and Feldman 1984; Klingemann 1979). They help citizens orient themselves to politics, organize their political beliefs, and process new political information.

With the rise in environmentalism, there is evidence that green orientations are becoming a source of identity with the Left or Right for many individuals. Robert Rohrschneider (1993a) maintains that environmental orientations are integrated into a general Left/Right schema for many Europeans (even if these orientations are still only weakly integrated into party alignments). Dieter Fuchs and Hans-Dieter Klingemann (1990) find that the citizens' interpretations of Left and Right change over time to reflect a growing interest in postmaterial issues. To an increasing extent, especially among the young, to be a leftist now implies a commitment to environmentalism and green political values. Similarly, Inglehart (1990, chap. 11) finds that Left/Right orientations are a strong predictor of support for environmental groups among the European public.

Finally, we must examine whether support for environmental groups is linked to general orientations toward the democratic process. Analysts have

MOVEMENTS IN CONTEXT

variously described the environmental movement as a vanguard for democratic reform or as an agent of revolutionary change. Indeed, there are elements of both within the movement. The more general question, however, is whether support for the movement is related to dissatisfaction with the democratic order, as critics of the movement sometimes claim. The Eurobarometer series includes data on system support, which we can use to determine whether political orientation is related to support for environmental interest groups.

The above variables will enable us to determine which factors are important in defining the base of nature protection and ecology groups and thus the political basis of the green movement.[14] Table 3.6 presents a multivariate model of potential membership (members as well as those willing to become members) in nature protection groups and ecology groups based upon the combined European sample from Eurobarometer 25. We use a behavioral measure to evaluate a deeper sense of support than mere approval of these groups. This analysis determines which of the above factors—social class, education, age, postmaterial values, Left/Right orientations, and satisfaction with democracy—have a strong and consistent impact on support.

One of the most obvious patterns is the general parallel in relationships for nature protection groups and ecology groups, even though there is generally a greater willingness to join nature protection groups. In terms of the relative importance of predictors, psychological factors have a stronger impact on support than do social location variables (see also Rohrschneider 1990). Postmaterial values emerge as the strongest single predictor of potential membership in ecology groups ($\beta = .16$) or nature conservation groups ($\beta = .16$). Nearly half (46 percent) of postmaterialists say they would potentially join an ecology group, whereas less than one-sixth (13 percent) of materialists express interest in possible membership. Left/Right orientations are also strongly related to support for green groups. Despite claims that the movement is neither Left nor Right, a comment echoed in our interviews with environmental group representatives, self-identified leftists are more likely than rightists to support environmental groups.

The data also indicate that the green movement has a distinct social base in the young and better-educated strata of society. More than a third of younger Europeans (34 percent) express a willingness to join ecology groups, and more than half (55 percent) express similar orientations toward nature protection groups. In contrast, less than one-sixth (14 percent) of the oldest age cohort are members or potential members of ecology groups. Bettereducated Europeans also display greater support for both nature protection and ecology groups. Indeed, if we combine these characteristics we have a

TABLE 3.6

Correlates of Potential Membership in Nature Protection and Ecology Groups for the European Public, 1986

Variable	Nature Protection Groups			Ecology Groups		
	Membership	Eta	Beta	Membership	Eta	Beta
Social class						
New middle class	53			32		
Old middle class	48			27		
Working class	43			23		
Farmers	36			18		
Retirees	36	.14	.07	17	.16	.09
Education						
Low	35			18		
Middle	46			23		
High	48			27		
Advanced	56			35		
Still a student	68	.18	.09	42	.21	.13
Age						
Under 30	55			34		
30–45	47			29		
46–60	41			22		
61 and over	34	.15	.07	14	.18	.11
Value priorities						
Materialist	30			13		
Mixed	47			25		
Postmaterialist	65	.22	.16	46	.23	.16
Political orientation						
Left	48			39		
Left/Center	51			36		
Center	46			24		
Right/Center	47			17		
Right	38			16		
No opinion	27	.13	.09	15	.18	.14

TABLE 3.6 (continued)

	Nature Protection			Ecology Groups		
	Membership	Eta	Beta	Membership	Eta	Beta
Satisfaction with democracy						
Satisfied	51			24		
Dissatisfied	40	.15	.11	27	.06	.03
Multiple R			.32			.33

Source: Eurobarometer 25 (May 1986).

Note: Figures are the results of a Multiple Classification Analysis predicting potential membership in nature protection and ecology groups. The columns list the unadjusted percentages of potential members for each category of the predictor variable. The eta coefficient is a measure of the simple correlation of each variable; the beta coefficient is a measure of the impact of each predictor independent of the other variables in the model. Respondents with missing data on the potential membership variable are excluded from these statistical models.

potent predictor of support for the environmental movement. Among better-educated younger Europeans, 43 percent express a willingness to join ecology groups. To a real extent, the ecology movement reflects the rise of a new generation of Europeans, nurtured in the affluence of the postwar economic miracle and imbued with values that build upon this new experience.

Another important result of the survey concerns the variables that do not display significant effects. For instance, despite the theoretical link between the rise of a new middle class and the emergence of NSMs, class differences in environmental support are weak. Thirty-two percent of the new middle class say they are potential members of ecology groups, but so, too, do 27 percent of the old middle class.[15] Overall, there is barely a 10 percent gap in the environmental orientations of the new middle class and the working class. Although the leadership cadre of the environmental movement and even its membership is heavily drawn from the new middle class (Cotgrove and Duff 1980; chap. 5, this vol.), the social base of environmentalism does not have this distinct class accent. Indeed, as a movement stressing popular collective goods, the reach of environmentalism across class boundaries may not be a surprising finding. The other significant non-finding is the general absence of a relationship between satisfaction with the democratic process and support for environmental groups. Among the European public as a whole, potential membership in ecology groups is only slightly higher among

TABLE 3.7

Correlates of Potential Membership in Nature Protection and Ecology Groups by Nation, 1986

	FRN	BEL	NET	GER	ITA	DEN	IRE	BRI	GRC
Nature protection groups									
Social class	.07	.18	.11	.11	.09	.10	.14	.09	.13
Education	.11	.17	.16	.12	.21	.08	.18	.12	.15
Age	.18	.03	.07	.14	.22	.13	.17	.11	.12
Value priorities	.07	.22	.13	.14	.09	.08	.05	.16	.14
Political orientation	.07	.15	.13	.15	.10	.11	.08	.06	.17
Satisfaction with democracy	.07	.22	.13	.14	.09	.08	.05	.16	.14
Multiple R	.30	.37	.33	.27	.39	.29	.36	.31	.43
Ecology groups									
Social class	.09	.26	.09	.13	.08	.13	.16	.11	.12
Education	.05	.18	.06	.04	.16	.15	.20	.16	.11
Age	.12	.11	.07	.21	.16	.12	.16	.08	.11
Value priorities	.08	.13	.11	.19	.11	.15	.07	.19	.16
Political orientation	.06	.18	.18	.26	.12	.14	.09	.10	.18
Satisfaction with democracy	.05	.14	.09	.11	.03	.03	.07	.05	.10
Multiple R	.23	.38	.28	.50	.34	.31	.38	.35	.42

Source: Eurobarometer 25 (May 1986).

Note: A Multiple Classification Analysis (MCA) was used to predict potential membership in nature protection and ecology groups in each nation. The figures are the beta coefficients from MCA, which measure the impact of each predictor independent of the other variables in the model. Respondents with missing data on the potential membership variable are excluded from these statistical models.

those who are dissatisfied with the functioning of democracy (27 percent) than among those who are satisfied (24 percent). For nature conservation groups, potential membership is actually higher among those satisfied with the democratic process. This reversed pattern of correlation provides partial evidence of the differing relationship that conservation groups and ecology groups have to the established political order.

We also analyzed the potential membership in these two types of groups based on country (table 3.7). The national analyses largely repeat the patterns found in the European analyses; for instance, age is consistently linked to support for environmental groups, and satisfaction with democracy is weakly related to potential group membership. In addition, however, there are several important national differences to note. Psychological factors, such as post-materialism and Left/Right orientations, have a stronger impact on group involvement in the Northern European, multiparty systems where green issues have been integrated into political competition. Some of the strongest relationships between Left/Right attitudes and participation in ecology groups are found in Germany (β = .26), the Netherlands (β = .18), Belgium (β = .18), and Denmark (β = .14). Conversely, educational skills are more important in predicting environmental activism in nations where the movement largely remained outside of party and Left/Right alignments at the time of our study, such as in Britain, Ireland, and Italy.[16]

On the whole, these results point to a general similarity in the social base of environmental action within Europe. The environmental movement draws heavily upon members of the postwar generation, especially better-educated youth. This is a movement of postmaterialist youth, who are disproportionately represented among the supporters and activists of the movement. The environmental movement also includes at least psychological ties to progressivism and the European Left; leftist support for environmentalism is especially clear for ecology groups. Among younger Europeans, environmental orientations are an important factor in determining broader political identities. Thus in a very real sense, public support for the environmental movement can be seen as a consequence of the affluence and security that transformed modern European societies and their citizens.

Europeans and the Environmental Movement

The evidence on the political context of the environmental movement is straightforward: there has been a greening of European public opinion. By large majorities, European publics express their support for the protection of the environment and the actions required to carry out environmental reform. Perhaps the depth of this sentiment is reflected in the public's endorsement of strong environmental protection measures, even if economic growth or inflation suffers as a consequence. Although one should realize that it is easy to voice such views in a public opinion survey, when real economic costs are hypothetical, there is also evidence that growing environmental consciousness is affecting consumer and electoral behavior.

The public opinion data also underscore the point that support for environmental reform is not a sunshine issue that exists only when other important economic and social problems are in abeyance. Popular support for environmental reform persisted through the economic downturns of the past decade. If anything, there has been a gradual trend toward greater environmental concern among Europeans since the 1970s, a trend that accelerated at the end of the 1980s. Public opinion data from the United States mirror this pattern (Dunlap 1992). To an increasing extent, environmentalism has become a firm element of the political agenda and political thinking in Western democracies.

These patterns of popular support for environmental reform and specific environmental groups hold several implications for the green movement. The breadth of support implies that the movement generally has a moderate, reformist image. If citizens thought that environmentalism would involve a revolutionary reshaping of social, economic, and political relations (as some environmentalists proclaim), then support for the movement would likely be more limited. Instead, we interpret these data as evidence that Europeans view the green movement as an agent of political reform, one that addresses hitherto undiscussed problems and the quality of life. This reformist image is consistent with the movement's image of itself.

Europeans do make some distinctions between the two elements of the environmental movement. Popular support for traditionally oriented conservation groups is virtually universal; these groups represent consensual social goals. The greater political demands made by ecology groups test Europeans' commitment to environmental reform, and support is therefore less common. Frankly, we doubt whether most Europeans would endorse the more vibrant shades of a green future proposed by radical environmentalists, such as Rudolf Bahro, Jutta Dittfurth, or Greg Foreman. Yet this should not detract from the evidence of the public's desire for change: even ecology groups receive positive approval ratings from most Europeans.

The breadth of support for a social movement should affect the political goals and activities of SMOs that draw upon this base. A social movement that begins with a popular base is obviously in a different strategic position than a movement representing an unknown or unloved minority. A weak or politically disfavored social movement may need to adopt confrontational or revolutionary methods to convert or overwhelm the majority. Movements representing political minorities, which is the norm, are thus vulnerable to counteractions by the dominant political forces. Such minorities' movements also have fewer resources to draw upon in developing their challenge to the political order because they represent a small sector of society, often a sector with marginal economic or political resources.

In contrast, the popular support for environmentalism gives the movement an advantageous position in mounting its political challenge. This support grants great legitimacy to environmental groups as representatives of the popular will; they can rightly claim to speak for the majority of Europeans who favor environmental reform, not just for a distinct political minority. Environmental groups should all be familiar with the statistics we have presented in this chapter, citing these data as evidence of the legitimacy of their basic policy claims. A supportive public also provides a large potential constituency for the movement. Thus, nature protection groups have been exceptionally successful in soliciting contributions and even in organizing joint fund-raising activities with business and other social groups. Ecology groups have also marshaled considerable resources through individual memberships and financial contributions. The entire movement has benefited because its core supporters—the young and better educated—are easily mobilized into political action. In short, environmental interest groups do not have to adopt revolutionary methods to convert or overwhelm the majority— the green movement need only mobilize the majority it can already claim. The basic challenge facing environmental interest groups is to mobilize and educate these existing predispositions rather than to convert opinion.

3

ENVIRONMENTAL

ORGANIZATIONS

Chapter Four

THE ORGANIZATION OF

ENVIRONMENTALISM

Public interest in environmental issues has grown to new heights over the past decade, projecting these issues onto the political agenda of most Western industrial democracies. Public opinion, however, provides only a latent base for political action, which must be mobilized, or at least coordinated, to affect the political process. Once a movement develops organized groups that direct the actions of the movement, it becomes important to study the behavior of these groups.

As we stated in chapters 1 and 2, much of the energy of the contemporary environmental movement is linked to politics through a network of interest groups and other members of the environmental lobby. Environmental groups are important because they provide an organized base for environmentalism. Policy influence in contemporary societies requires the formal representation of interests. Legislation must be monitored, close contact must be maintained with the bureaucracy, authoritative witnesses must be available to present environmental views, and the organization must be able to educate and mobilize its supporters. In pursuing these activities environmental groups are defining the agenda and political direction of the movement. These groups provide the public with cues as to which policies are important and which strategies should be adopted.

In this chapter we examine the basic organizational characteristics and structural features of the major environmental interest groups in Western Europe. This factual information provides valuable evidence on several theoretical issues as well. For instance, by dating the origins of environmental groups we can judge the developmental pattern of the environmental movement and its changing composition over time. An examination of the political resources controlled by these groups (their membership, financial base, and personnel) highlights the value and limits of resource mobilization theory. And finally, by examining the internal structure of these organizations we can evaluate the often-cited claim that environmental groups, and other NSMs, violate one of the central theories of political organization—the iron law of oligarchy. The findings introduce the groups we surveyed and describe the diversity in the contemporary environmental movement.

Environmental Organizations

In studying European environmental groups we wanted to capture the various shades of the green rainbow—from the traditionalism of the Royal Society for the Protection of Birds (RSPB) to the assertive tactics and alternative political goals of Robinwood. In addition, we wanted to represent the organizational diversity of the movement. Although we are primarily interested in the political activities of mass-membership groups, resource mobilization theorists would describe environmentalism as a social movement industry that interconnects mass-membership groups, research institutes, educational foundations, political lobbies, and small associations of environmental elites. A comprehensive study of the environmental movement should include all of these elements.

We also wanted to tap the different ideological strands of environmentalism to examine the impact of ideology on the structure and behavior of these organizations. As we discussed in preceding chapters, the major ideological cleavage that exists within the environmental movement distinguishes between groups with a conservationist or ecological perspective. Conservation groups are predominantly concerned with issues of wildlife preservation and the protection of natural resources. In pursuing these goals, moreover, conservation groups generally accept the existing norms of the socioeconomic system and work within this value structure. In contrast, ecology groups are more likely to focus their attention on the emerging problems of advanced industrial societies (nuclear power, industrial pollution, protection of the quality of life, and the development of renewal resources). In addressing these issues, ecology groups should identify the existing socioeconomic system as a source of these problems and advocate an alternative value system or social structures as part of the solution. The key factor in distinguishing between these two types of organizations is their acceptance of the prevailing social paradigm or their adherence to the New Environmental Paradigm.

We used several criteria in selecting groups for this study (see appendix for information on the group selection criteria and methods). Our overarching objective was to include the five or ten major environmental organizations in each nation, as well as to represent the major political forces within each national environmental network. After collecting preliminary information on such groups in ten European states, we conducted interviews with the representatives of sixty-nine organizations (table 4.1).

To examine the role of ideology among environmental groups, we classified each organization in terms of its conservation or ecological orientation, based on two criteria. First, we consulted other publications that have categorized European environmental groups along a similar continuum (Cotgrove 1982; Lowe and Goyder 1983; Milbrath 1984; Rucht 1989; Diani 1988, 1990a; Vadrot 1978). Second, we used the information from the personal interviews, supplemented by the published materials of the organization. We felt it best to make a global evaluation of conservation or ecological identity, rather than use one question in our survey as the basis of classification, because we are trying to establish a general orientation that transcends any specific question and because this external classification ties these groups to their status in prior research and theorizing. Even if such a dichotomy is sometimes forced—in that any organization contains some of both orientations—these designations reflect the primary dimension of ideological differentiation existing within the movement.

We included most of the major nature conservation groups in Western Europe in our survey (denoted by a plus sign before their names). For instance, we interviewed representatives of bird societies in each country where a large national organization exists (the one omission was France). Other conservation groups ranged from most national branches of the WWF to influential nature protection groups such as the Danish Association for Nature Conservation (DN), An Taisce of Ireland, and the Council for the Protection of Rural England (CPRE).

Forty-eight of the organizations listed are mass-membership groups, many of which advocate an ecological perspective (indicated by an asterisk). We surveyed nearly all of the active affiliates of Friends of the Earth and Greenpeace within the European Community. NOAH, in Denmark, and Robinwood, in West Germany, are examples of loosely structured, youth-oriented groups that assertively challenge the prevailing social norms; the Conservation Society, in Britain, and the Foundation for Alternative Living, in Holland, similarly advocate changes in life-style and the creation of a sustainable society as necessary steps in addressing contemporary environmental problems.

TABLE 4.1

European Environmental Groups Interviewed for This Study

BELGIUM
Bond Beter Leefmilieu
*Friends of the Earth (AdlT)
*Greenpeace
+Inter-environnement Wallonie
+National Union for Conservation
Raad Leefmilieu te Brussel
+Réserves Ornithologiques
+Stichting Leefmilieu
+World Wildlife Fund

DENMARK
+Danish Association for Nature Conservation (DN)
+Danish Ornithological Association (DOF)
+Friluftradet
+GENDAN
*Greenpeace
*NOAH

FRANCE
+COLINE
CREPAN
+French Federation of Societies for the Protection of
    Nature (FFSPN)
*Friends of the Earth (AdlT)
*Greenpeace
+Institute for European Environmental Policy (IPEE)
+Journalists and the Environment
Nature and Progress
+Society for the Protection of Nature
+World Wildlilfe Fund (WWF)

GREAT BRITAIN
+Civic Trust
+Conservation Society
+Council for Environmental Conservation (CoEnCo)
*Council for the Protection of Rural England (CPRE)
+Fauna and Flora Preservation Society (FFPS)
*Friends of the Earth (FoE)
Green Alliance
*Greenpeace
+Royal Society for the Protection of Birds (RSPB)
Town and Country Planning Association (TCPA)

GREECE
Ellinike Etairia
+EREYA
*Friends of the Earth
+Friends of the Trees
+Hellenic Society for the Protection of Nature (HSPN)
PAKOE

IRELAND
+An Taisce
+Wildlife Federation

ITALY
+Agriturist
*Friends of the Earth (AdT)
+Fund for the Italian Environment (FAI)
+Italian League for Bird Protection (LIPU)
+Italia Nostra
*League for the Abolution of Hunting
*League for the Environment–Arci
+World Wildlife Fund

LUXEMBOURG
*Mouvement Ecologique
+Natura

NETHERLANDS
Association for the Protection of the Waddenzee
+Dutch Bird Protection Association
*Foundation for Alternative Living
*Foundation for Environmental Education (SME)
+Foundation for Nature and the Environment (SNM)
*Greenpeace
+Institute for Nature Conservation Education (IVN)
*Society for the Defense of the Environment (VMD)

WEST GERMANY
*Federal Association of Citizen Action Groups for
    Environmental Protection (BBU)
German Federation for Environmental and Nature
    Protection (BUND)
+German Federation for Bird Protection (DBV)
+German Nature Protection Ring (DNR)
*Greenpeace
*Robinwood
+Schutzgemeinschaft Deutscher Wald
+World Wildlife Fund (WWF)

Note: Conservation groups are marked by a plus sign; ecologist groups by an asterisk.

National federations of environmental groups constitute another component of this study (twelve organizations). Most of these federations unite a variety of local and single-interest groups under one national umbrella organization. The Federal Association for Citizen Action Groups for Environmental Protection (Bundesverband Bürgerinitiativen Umweltschutz, or BBU) was formed in the early 1970s as an association of local citizen action groups in West Germany and advocates an ecological perspective. Most umbrella organizations, however, adopt moderate or conservationist views because of the nature of their membership. The German Nature Protection Ring (DNR) and the Friluftradet in Denmark bring together groups concerned with outdoor activities, sports, and nature issues. The French Federation of Societies for the Protection of Nature (FFSPN) includes over one hundred French groups, ranging from the National Society for the Protection of Nature (SNPN) to Greenpeace. Separate regional organizations exist in Belgium (Inter-Environnement in Wallonia, Bond Beter Leefmilieu in Flanders, and the Raad Leefmilieu te Brussel in Brussels). Some federations were created as an institutional channel between the government and the environmental movement: the Dutch Foundation for Nature and the Environment (Stichting Natuur en Milieu, or SNM) and the British Council for Environmental Conservation (CoEnCo) are prime examples of this. Less formal associations exist in other nations.[1]

Part of the growing strength of the environmental movement is the support it receives from a network of environment research institutes and educational foundations. The Institute for European Environmental Policy (IPEE), for instance, maintains offices in Paris, London, and Bonn.[2] The separate IPEE offices conduct policy research for government agencies and are involved in efforts to educate the public on environmental issues. PAKOE performs similar activities in Greece, as does the Stichting Leefmilieu in Belgium. (The Öko-Institut is a major research and educational resource of the German ecological movement, though it is not included in this study.) The existence of separate conservation and ecological networks is illustrated by the dual system of environmental education in Holland: the larger and more established Institute for Nature Conservation Education (Institut voor Natuurbeschermingseducatie, or IVN) focuses its public education efforts on nature conservation issues and presents a traditional view of these issues; the smaller Foundation for Environmental Education (SME) was formed by the ecological Actiongroup Straw to present an alternative view of environmental issues.

Finally, a few groups of environmental elites facilitate communication among social leaders who are actively interested in environmental matters.

The Green Alliance in Britain and COLINE (Comité Législatif d'Information Ecologique) in France encourage a concern for environmental issues among the few hundred social and political elites who are invited members. Other organizations focus on a specific profession, such as the scientifically oriented EREYA in Greece or the Association of Journalists and the Environment in France.

Overall, our study contains thirty-nine groups with a traditional conservation orientation and twenty-one ecology groups (we coded the remaining nine groups as having mixed orientations). This classification of organizations agrees substantially with other published studies.

As we discussed in chapter 2, this distinction between conservation and ecology orientations partially transcends issue interests. Most wildlife protection groups, for example, accept prevailing sociopolitical norms and are categorized as conservationist organizations, but Greenpeace is extremely active in wildlife protection and still challenges the prevailing social system as part of its efforts to address wildlife problems. Conversely, other groups that advocate social change, such as macrobiotic agricultural groups, are asking for technological reform rather than challenging the social order. Environmental orientations also cut across organizational structures. Mass-membership groups obviously differ in their conservation or ecological orientation, but so do federations and research institutes. The BBU is an active proponent of alternative politics in Germany, although most other federations are more conventional in their viewpoints. Some research and educational institutes express a conservationist viewpoint (IPEE and IVN), whereas others (SME and the Öko-Institut) provide a resource base for ecologists.

The purpose of this research is twofold. First, we describe the organizations that channel the energies of the environmental movement and influence the political process of Western industrial democracies. Rather than a sample, this list comes close to defining the universe of major national environmental interest groups, providing a unique and valuable research base. Second, we have argued that the ideological orientation of a political organization can broadly affect its structure, goals, and tactics. Through a systematic comparison of conservation and ecology groups, we are able to test this theory of ISA.

Group Origins

The historical and political origins of national groups are central to the study of the environmental movement. How long have environmental groups been

FIGURE 4.1
Year of Formation of Environmental Groups   *Source:* See table in appendix.

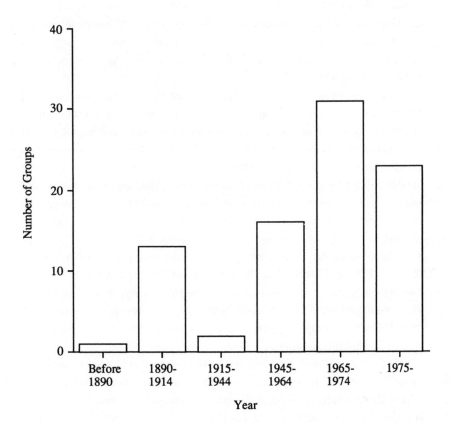

active in Western Europe, and how has the composition of the movement changed over time?

Traces of the two mobilization waves—the first during the 1800s, and the second during the 1960s and 1970s—are clearly visible in dating the origins of the groups in our study (fig. 4.1).[3] The initial peak of group formation took place between 1890 and the outbreak of World War I. Of course, many other environmental groups established during this period have subsequently dissolved or merged with other groups, but over a dozen organizations in our study had their origins in the period 1890–1910. Only one contemporary environmental group, the French SNPN, was established before 1890.

As Lowe and Goyder (1983, 16) found for the British environmental movement, overall organization-building efforts in Europe peaked in the turbulent decade between 1965 and 1975. Indeed, the largest portion of the contemporary movement originated during this period. Although this mobilization wave crested in the early 1970s, new environmental organizations are still being formed. Environmentalists created nearly two dozen important national groups after 1975.

The separate waves of environmental mobilization not only increased the number of organizations but also restructured the ideological orientations of the movement. The earlier one established the conservation movement through the creation of the first national bird societies and nature conservation groups; the second reinvigorated the existing conservation groups and added another layer of organization (such as the World Wildlife Fund, FFSPN, and DNR). Conservation groups thus span both mobilization waves; the earliest of the conservation groups in our sample was formed in 1854 and the most recent one in 1979 (the average year of formation is 1947).

The wave of the 1960s went beyond conservationism to institutionalize the new ecological approach to the environment. During this period FoE and Greenpeace groups, as well as other alternative ecology groups, appeared throughout Europe. All of the ecology groups we surveyed were established after 1965 (the average is 1976)! Ecologists are the new environmentalists, at least in chronological terms.

The composition of the environmental movement thus has changed dramatically over the past few decades. From the late 1800s until the mid-1960s, the movement was virtually synonymous with the protection of wildlife and the preservation of nature. Beginning in the 1960s, a new set of ecology groups emerged—groups concerned with the problems of advanced industrial societies and willing to challenge the goals and tactics of society in addressing these problems—adding a new, vibrant shade of green to the movement. As we compare the characteristics of conservation and ecology groups in the following chapters, we can determine the significance of this compositional change in the contemporary movement.

Organizational Resources

Research mobilization (RM) theory draws our attention to the importance of acquiring and managing resources for the pursuit of organizational objectives (McCarthy and Zald 1977; Zald and McCarthy 1987). Most organizations seek to survive not only for the continuation of the organization but also for the continued pursuit of its goals. RM maintains that the accumulation of resources is a crucial measure of the development of a social movement. Moreover, RM theorists argue that this mobilization of resources is a central activity of most social organizations, sometimes even displacing the primary social and political goals.

The importance of mobilizing resources for organizational maintenance and success is undeniable, but the RM approach misses the equally important interaction that can occur between the political identity of an organization and its potential for resource mobilization. Our framework of ISA suggests that the environmental orientation of a group should affect both the type of resources that it requires and the type of resources it can most easily mobilize. For instance, an established conservation group can turn to society's elite as well as to government offices for funding; a group such as NOAH or Robinwood does not have this option when campaigning. Even an almost universally valued resource—such as a group's reputation—varies in the traits (and reference group) that constitute a positive reputation. The positive reputations of Greenpeace and the RSPB represent distinctly different characteristics that are valued only within specific contexts. All groups mobilize resources, but possibly through different channels and of different qualities.

This section examines the basic organizational resources commanded by environmental interest groups—membership, finances, and personnel—and considers how the nature of an organization interacts with its process of resource mobilization.

Membership

For many groups, their primary resource is their membership base. Members provide a significant potential source of revenue. Group representatives openly acknowledged this fact during our interviews, as they discussed their strategies for increasing their organization's financial base through membership drives or the further solicitation of present members. This financial imperative is evident across the entire environmental spectrum. Many groups also mobilize the volunteer work of members in support of the group's activities. Most membership activities are politically neutral, as when members of a bird protection society conduct a bird census, but the membership also can furnish a potential pool of political activists. Campaigning groups, for example, rely upon their members as a source for mobilizing a protest or petition campaign (see, e.g., Klandermans et al. 1988). Conventional groups also turn to their members when organizing letter-writing campaigns to the government and media.

For public interest organizations, such as environmental groups, their membership base has additional meaning. A large membership provides legitimacy for organizations that claim to speak for the public interest. One British environmentalist noted that government ministers are impressed by the large membership of his organization and the potential political force the membership represents: "For every member of our organization, politicians assume there are several other voters who share these views. This makes them stop to think about what we are saying." The claim of representing its members provided this group with an entree to government offices and positions on policy advisory committees. Just as openly, the head of another organization with a dwindling membership lamented the loss of influence that has accompanied this decline; he acknowledged that another, more dynamic organization had taken over its place on many policy forums. Similar accounts of the importance of numbers emerged in other interviews, from both large and small organizations. Only a very few groups—most often antiestablishment ecology groups—feel their effectiveness is linked to their small size. In general, however, people count numbers, and the number of members is an important measure of the ascribed importance of an environmental group.

Mass-membership groups come in a variety of sizes (fig. 4.2; see the appendix for the membership of specific groups).[4] In the mid-1980s, the average group comprised about ten thousand members, but the dispersion around this figure is considerable. Among the groups represented in the figure, nearly one-third had fewer than five thousand members. At the other end of the spectrum, only seven organizations—all conservation groups—have a membership of one hundred thousand or more: the British National Trust, the RSPB, the Royal Society for Nature Conservation (RSNC), the German Federation for Bird Protection (DBV), the Danish Association for Nature Conservation, the Dutch Association for the Preservation of Natural Monuments, and the Dutch WWF. As discussed in chapter 3, European publics are more supportive of and more likely to join conservation groups because of their consensual goals; the average membership of conservation groups (42,000) outstrips that of ecology groups (22,000).

One of the hotly debated issues in research on social movements concerns the motivations that lead individuals to join social groups (e.g., Olson 1965; Mitchell 1979; Klandermans et al. 1988). Because we surveyed the organizations themselves, rather than their members, our evidence on this topic is indirect, though still relevant. We presented a list of various motivations to the representatives of mass-membership groups and asked them to rank the reasons they thought people joined their organization (table 4.2).[5] Although

ENVIRONMENTAL ORGANIZATIONS

FIGURE 4.2

Membership Size of Environmental Groups   *Source:* See table in appendix.

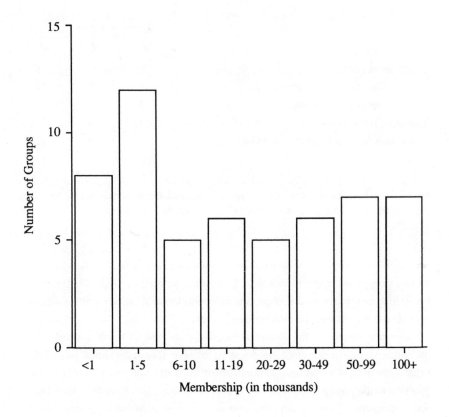

individuals may have several reasons for joining environmental groups, most officials feel that their members are motivated primarily by efforts to register support for the goals of the group; 75 percent of the groups rank this as the most important reason for joining. The chairperson of one Greenpeace organization related that people join the organization because they like the sight of rubber boats in front of whalers and Greenpeace activists protecting baby seals; the director of a bird protection group discussed his membership's love of birds. In sharp contrast, material incentives—access to the special benefits and privileges of membership—are seen as the least important motivation (ranked first or second by only 14 percent). In between these two extremes are motivations such as participating in voluntary work and seeking social contact. The emphasis on purposive goals among European environ-

TABLE 4.2

Perceived Motivations for Joining an Environmental Group

	Ranking, as Percentage		
	1st	2d	3d
To register support for group aims	75	11	5
To further personal point of view and influence society	16	34	34
To seek social contact and companionship	7	23	30
To get involved in volunteer work	7	34	25
To receive special benefits and privileges of membership	7	7	23
Total	112%	109%	117%

Note: Responses are from representatives of 44 mass-membership groups. Group officials listed the top three reasons why they thought individuals joined their organization; tied rankings are possible, so that totals may exceed 100 percent.

mental groups is nearly identical to Lowe and Goyder's (1983, 39) findings for British environmental groups, though continental groups ascribe a markedly lower importance to material incentives.

We are aware of the potential biases in the use of such a projective question. Group representatives want to believe that individuals join their organization as a conscious expression of support, not just to receive the glossy monthly magazine that the group produces. To claim that members are a measure of policy support also adds to the political legitimacy of a group. Still, the supporting evidence we collected tends to substantiate the pattern of membership motivations expressed by the group officials. Almost all environmental groups provide some direct product or benefit to their members, such as a membership magazine or educational opportunities, but very few groups are able to provide sufficient material (or selective) benefits to offset the financial costs of membership.[6] The "product" of group efforts is normally collective and nonmaterial goods that are not subject to selective distribution. The glossy magazine or occasional membership newsletter may remind individuals why they joined an organization, but such incentives by themselves are not the primary motivation or payoff. Moreover, the modest cost of membership in most groups (averaging about fifteen dollars a year) is unlikely to evoke narrow, self-interested calculations among individuals interested in environmental issues (for similar findings from American public interest groups see Berry 1977, chap. 2; Schlozman and Tierney 1986, chap. 6). Mancur Olson's *Logic of Collective Action* may explain why some pro-environment citizens choose not to join groups; but those who do join, almost by definition, are not governed by calculations of direct benefits. In addition, we shall show that the limited opportunities for face-to-face interaction in most environmental interest groups limits the number of individuals who are motivated by solidarity and participatory benefits. In sum, membership in most groups seems to represent an expression of support for the goals and objectives of the organization, and the public relations activities of most groups underscore their reliance on such purposive goals.

Returning to the question of group size, we find that the environmental movement is fragmented into many distinct national groups of modest size (and a larger number of small single-interest and local organizations). This fragmentation might explain the existence of federations in most nations. In uniting more than a hundred smaller groups, for example, the FFSPN indirectly represents a combined membership of about 850,000 French citizens; the largest French mass-membership group in our study has fewer than 20,000 members. Similarly, the groups belonging to the German DNR boast a membership of 3.3 million.[7] The three regional federations in Belgium are primary representatives of the larger environmental movement in matters of national policy making.

When we combine the mass-membership and umbrella organizations, the membership in European environmental groups is considerable. Although precise figures are difficult to calculate, our best efforts suggest that about ten million Europeans belong to the organizations included within this study.[8] These figures, along with the measures of popular support for environmental policies discussed in chapter 3, strengthen the basic political legitimacy and influence of the movement.

The growing size of the environmental movement is as impressive as its absolute membership levels. Over half of the groups we surveyed report that their membership had increased by 25 percent or more over the past five years; only 10 percent report an equivalent decrease. The general growth in membership also cuts across ideological lines. Many large conservation groups posted large increases in membership during the early 1980s: the RSPB expanded by about 50,000 members, WWF in Germany grew from 5,000 to 40,000 members, and the Danish Association for Nature Conservation attracted almost 100,000 new members over the same time span. The majority of ecological organizations, though not all, shared in this expansion.

Significantly, our accountings predate the Chernobyl disaster, which analysts widely credit with stimulating environmental activism. Several recent surveys of environmental groups in a single nation report dramatic

TABLE 4.3

National Differences in Membership of Major Environmental Groups

Nation	Greenpeace	Friends of the Earth	World Wildlife Fund	Ornitho-logical Society	Total
Britain	55,000	27,000	80,000	407,000	569,000
West Germany	76,000	0	40,000	140,000	256,000
Netherlands	80,000	15,000	100,000	30,000	225,000
Denmark	45,000	0	50,000	9,000	104,000
Italy	0	1,000	62,000	23,000	86,000
Belgium	6,000	1,000	25,000	18,000	50,000
France	7,000	3,000	15,000	5,000	30,000
Greece	0	1,000	0	0	1,000
Ireland	0	0	0	0	0

Note: Data on Luxembourg were not available.

growths in membership in the late 1980s. Gene Frankland (1990) reports that from 1985 to 1989 FoE-Britain grew from 27,000 members to 125,000 and Greenpeace increased its donors from 55,000 to 281,000. British membership figures for 1992 list a membership of 2.3 million for the National Trust, 839,000 for the RSPB, 411,000 for Greenpeace, 240,000 for FoE, and 230,000 for WWF (*Economist,* June 6, 1992, 63). Mario Diani (1990a) reports equally impressive gains for WWF-Italy and the League for the Environment; Jacqueline Cramer (Jamison et al. 1990) describes a similar growth in Greenpeace-Holland and other Dutch groups. Although one frequently hears that the movement is in decline, just the opposite is the case—environmentalism is probably the largest public interest movement in Western Europe, and its membership base expanded through the 1980s.

The considerable vitality of environmental interest groups does not mean that all nations are sharing equally in this development. Indeed, there is dramatic cross-national variation in the popular support and membership base of these groups. Table 4.3 illustrates the general pattern of these national differences by comparing the membership of four key environmental groups that exist in most nations: Greenpeace, FoE, WWF, and the national ornithological society. The combined membership of these organizations reaches substantial levels in Britain (569,000) and Germany (256,000). When one takes population size into account, the Dutch and Danish environmental movements have an exceptionally large ratio of members. The Dutch green movement is one of the most successful and well-organized in Europe; it has the largest Greenpeace organization, the largest WWF affiliate, and the second largest FoE affiliate in Europe. The combined membership of the Dutch groups nearly equals that of Germany, for example, although the total German population is nearly four times larger. The Danish movement has achieved similar success in developing its membership base. Moreover, the Danish statistics do not include the membership of the Danish Association for Nature Conservation, which boasts that one of every ten Danish households includes a member of DN!

The French movement probably represents the nadir of institution building. With the possible exception of the FFSPN, the French movement has been unable to develop a strong infrastructure for environmental action; the French bird society and WWF are the smallest in Europe, and French ecology groups are equally weak. This undeveloped infrastructure carries over to the French ecology party, which sputters between periods of vitality and morbidity. Symptomatic of these organizational problems, the French Greenpeace affiliate actually lost members after the bombing of the *Rainbow Warrior* in New Zealand by agents of the French government, although all other Greenpeace organizations in Europe grew substantially in reaction to the event. The Greek and Irish movements are (predictably) even smaller, but the relative weakness of the French movement still stands out as a dramatic exception to the overall advance of environmentalism in Europe.

We are skeptical of efforts to explain these cross-national differences in the strength of green parties or the environmental movement with national attribute data, such as income or the structure of the labor force (e.g., Kitschelt 1988, chap. 1; Müller-Rommel 1989). Too many forces are at work to produce systematic patterns across this small number of nations. Although national affluence is positively related to overall membership in environmental groups, this relationship is driven by results from Greece and Ireland. Among the other European states, there is little relationship based on level of affluence. Other national characteristics are not significantly related to membership, partially because we are analyzing only nine nations.[9] Even affluence, we would argue, is an indirect measure of more relevant concepts, such as whether economic needs are being fulfilled or where the unintended consequences of economic growth may be most extensive. Within each nation, the strength of the movement reflects a complex combination of systematic cross-national forces and idiosyncratic national conditions. The weakness of the French movement illustrates this complexity. One can trace the disarray among French environmentalists to organizational problems within the movement itself, the government's hostility toward environmental groups, and the public's general aversion to group-structured political activities (Wilson 1987, 1990; Kitschelt 1990; Vadrot 1978; DeClair 1986). Affluence has not produced a large movement in France. In contrast, in Denmark and Holland the successful growth of environmental groups is facilitated by a rich organizational life and a political culture that accepts diversity. Cultural forces also affect the nature of the movement; in Britain, popular support for environmental action generally has followed a course of moderation typical of British political culture, whereas German ecology groups are more ideological and politicized because of roots in the student and alternative movements.

Financial Resources

One of the most precious resources for environmental groups is money, as anecdotal discussions during our interviews amply demonstrated. Economic interest groups normally can depend on the solid financial support of business or labor, depending on their orientation, but financial needs are a persistent problem for public interest groups. Environmental groups lack a captive base of financial support, and the costs of challenging businesses, agricultural interests, and their entrenched representatives are considerable. Moreover, money has a special value because it is convertible into other needed goods: staff, advertising, consultants, public relations programs, and so forth. As American research has shown, sufficient financial support is often the greatest resource need of public interest groups (Schlozman and Tierney 1986, chap. 5).

In examining the annual budgets for European environmental groups, we calculated a normal operating budget, excluding most government contracts, in-kind support, and similar nonrecurring sources of income (fig. 4.3).[10] As is the case with membership statistics, the financial resources of environmental groups vary greatly. About one-fourth of all groups function with an annual budget of less than fifty thousand dollars. Often, these are the smaller ecology groups—such as Robinwood, the Foundation for Alternative Living, NOAH, or the Conservation Society—which depend on extensive support from volunteers and workers receiving unemployment compensation. At the other extreme, the top quartile of groups have annual budgets in excess of five hundred thousand dollars and are drawn from all sectors of the environmental movement: preservation associations (Italia Nos-

FIGURE 4.3

Operating Budget of European Environmental Groups, 1985   *Source:* Responses
from representatives of 69 environmental groups.

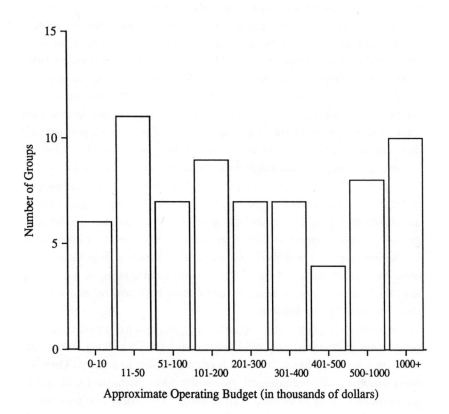

When these budgetary figures are combined, we conservatively estimate
that these groups control well over $50 million a year—small by the standards
of economic interest associations but large by the standards of European

tra, the British National Trust), traditional conservation groups (DN, the
Association for the Preservation of Natural Monuments), wildlife protection
groups (RSPB, the French League for Bird Protection, and WWF affiliates in
Britain, Germany, and Italy), and ecology groups (FoE in Britain, League
for the Environment in Italy, and Greenpeace affiliates in Britain, Holland,
and Germany). The diversity among these well-funded groups illustrates the
many elements of the movement that are able to mobilize support for their
activities.

When these budgetary figures are combined, we conservatively estimate
that these groups control well over $50 million a year—small by the standards
of economic interest associations but large by the standards of European

public interest groups. Furthermore, two-thirds of the groups report that even after adjusting for inflation, their incomes have increased over the past two or three years (see also McCormick 1991, 156). Finances thus furnish another indicator of the continuing expansion of the movement.

The sources of revenue are as important as the amount of revenue because revenue sources can affect the financial and political autonomy of an organization (Lowe and Goyder 1983, 43–44; Schlozman and Tierney 1986; Walker 1991, chap. 6). Dependence upon individual donations, for example, should make environmental groups more responsive to the interests of their supporters. At the same time, a dependence on membership income may leave groups financially vulnerable as the tides of popular support ebb and flow. On the positive side, one mass-membership group increased its income by almost $2 million in a single year through a special television appeal. Some groups, however, have experienced sharp drops in individual support that threatened the survival of the organization. Groups emphasizing individual donations might also undertake activities based on the likelihood of garnering public support instead of on intrinsic environmental importance. In addition, servicing and recruiting members demands a large share of the organizational resources of many mass-membership groups. A reliance on individuals as a source of revenue thus carries both positive and negative implications for group behavior.

In contrast to individual support, large institutional sources, such as government or business, may offer substantial financial backing. When provided on a recurring basis, these funds impart stability to a group. There are many examples of environmental groups that have established such ties to corporations seeking to develop a green image. Dependence on government grants carries the same benefits of a large-scale funding opportunity, as well as the possibility of recurring funding. These revenue sources, however, also involve potential costs in autonomy and reputation. More than once we have heard environmentalists repeat the proverb "He who pays the piper gets to call the tune."

Rather than ask for specific percentages, which requires more precision than most people can normally recall in an interview, we asked group representatives to rank the importance of various financial sources for their organization (table 4.4). The populist base of the movement is illustrated in the high level of individual funding. Nearly half of the organizational representatives (49 percent) specify membership dues as their primary source of income, with gifts and endowments from individuals ranked a close second.[11] Many groups are adept at marketing group merchandise, making this the third-ranked source of revenue. Greenpeace and FoE groups provide glossy

TABLE 4.4

Sources of Income for Environmental Groups

	Ranking, as Percentage			
	1st	2d	3d	No Support
Membership dues	49	18	9	6
Gifts or endowments from individuals	12	30	15	30
Sales of group materials, fund-raising activities, etc.	9	24	21	27
Grants from the central government	18	13	8	55
Grants from private trusts or foundations	2	6	16	58
Grants from other government agencies (e.g., local)	6	5	6	72
Donations from businesses	3	2	5	72
Income from investments	2	5	6	78
Total	101%	103%	86%	

Note: Responses from representatives of 67 organizations. Tied rankings are possible, so that totals may exceed 100 percent.

color catalogs of T-shirts, posters, buttons, and similar paraphernalia that are emblazoned with the group's logo and campaign slogans. As another well-known example, the Panda House shops of WWF sell stuffed animals, wildlife books and maps, and an incredible array of items imprinted with their panda mascot. In fact, many mass-membership groups included retail catalogs with the published material we initially requested. Marketing has become an important source of environmental funding.[12]

Several environmental organizations receive a significant portion of their revenues from governmental sources, primarily grants from the central government. The educational and research institutes are most dependent on government support, but several federations and mass-membership groups also rely on continuing government grants. If contract income and other government support were included in budget statistics, the importance of governmental sources would loom even larger. Still, many group representatives express caution about the possible co-optation that can accompany government grants, and half claim to receive no support from these sources. Support from business probably carries the greatest potential for restricting the autonomy of an environmental group or calling its credibility into question. Not surprisingly, only 18 percent of all organizations say they receive any support from business interests.

TABLE 4.5
Ranking of Income Sources

	Conservation		Ecology		
	1st/2d Rank	No Support	1st/2d Rank	No Support	Tau-c
Membership dues	74%	0%	100%	0%	.06
Gifts or endowments from individuals	62	9	47	18	−.06
Sales of group materials, fund-raising activities, etc.	18	48	41	12	.37*
Grants from the central government	22	61	0	88	−.22*
Grants from private trusts or foundations	9	57	6	71	−.11
Grants from other government agencies (e.g., local)	4	74	12	77	.02
Donations from businesses	8	61	0	82	−.18*
Income from investments	8	61	6	88	−.19*

Note: Responses are from representatives of 46 mass-membership organizations. Multiple responses are possible, so that percentages do not total 100 percent. Tau-c correlations significant at the .10 level are indicated by an asterisk.

A comparison of funding sources illustrates the importance of environmental orientations in the process of resource mobilization. Conservation and ecology groups are both concerned with environmental problems and derive their legitimacy through their memberships, yet their environmental orientations should still affect their resource base. As representatives of populist, antiestablishment forces, we expect ecology groups to rely more heavily on membership dues, the sale of group materials, and other individual sources of income. Conversely, conservation groups do not challenge the established social forces and thus should be more likely to receive support from governmental sources, private foundations, and even business interests.

The results of this comparison largely confirmed our expectations (table 4.5). The funding sources of conservation groups reflect their establishment orientations; they are more likely to rely upon government grants (tau-c correlation = −.22), donations from business (−.18), and possess large endowments that generate investment income (−.19).[13] Ecology groups, in contrast, place greater reliance on the sale of group merchandise (.37). All mass-membership groups heavily rely on membership dues, so there are only weak differences (.06) between conservation and ecology groups on this funding source (though in the expected direction).

It is difficult to draw causal inferences from the correlations presented in the table. These patterns of funding probably represent the joint decisions of environmental groups and potential funding sources, and one cannot sort out the direction of causality or separate the actions of donor and recipient. Organizations focus their fund-raising efforts on the sources they expect will be most supportive of their activities, and, likewise, funding agencies are more receptive to organizations that share their orientations. Although a traditional conservation group will approach a philanthropic foundation for support, Greenpeace will approach the general public with mass mailings on its campaigns. Similarly, a philanthropic foundation that has an interest in conservation issues is unlikely to approach Greenpeace to act as their agent. Thus, the different funding patterns reflect the networks of alliances and patterns of social support that exist for these two sets of environmental groups.

Staff Support

A third important resource is the staff support to carry out the everyday work of the organization: overseeing administration, serving members, mobilizing support, and representing the group to public and political officials. The existence of a full-time professional staff marks a crucial threshold for an organization, providing a continuity that enables the group to compete in the long process of policy formation. Indeed, the existence of such an institutional base was one factor leading us to study environmental groups as representatives of the movement. The skills of the staff are also vital to the success of an organization; seldom can a group rise above the abilities of its personnel.

As with other resources, the number of personnel may vary greatly. On average, most groups function with a small professional staff. Among the organizations we surveyed, the modal number of full-time employees is six. Even if we add part-time workers, the employee base of most groups remains quite small (fig. 4.4, top panel).[14] At the low end of this continuum, about a tenth of all organizations exist without any paid employees; volunteerism sustains the organization. At the other extreme, only one-sixth maintain a professional staff of more than twenty. Of the groups surveyed, RSPB has the largest professional staff, with more than forty full-time employees. Most organizations also utilize volunteers. Although volunteer workers are sometimes central to policy making and operation, as they are in NOAH and Robinwood, more often they perform maintenance functions, such as serving the membership, handling correspondence, and staffing an office.

FIGURE 4.4

Staff Size of Environmental Groups   *Source:* Responses from representatives of 69 environmental groups. Paid employees (FTE) in the top half of the figure include both full-time and part-time employees.

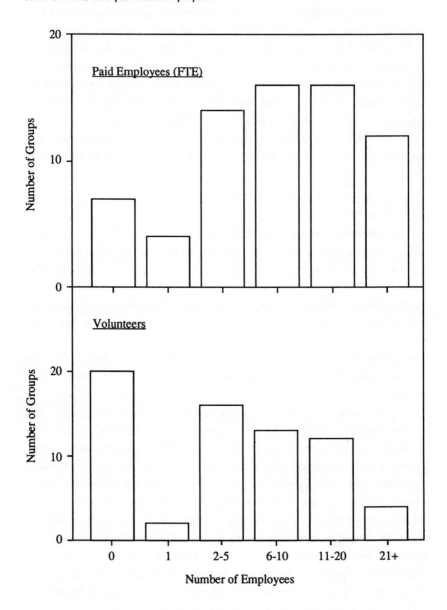

Nearly all organizations attempt to increase their staff (or mobilize more resources), but we find that a group's environmental orientations influence its choice of personnel.[15] Conservation groups are more successful in maintaining a large professional staff (tau-c $= .36$), retaining nearly twice as many employees as ecology groups. The average ecology group, conversely, has nearly twice as many volunteers as the average conservation group (.46). In fact, ecology groups are more than three times more likely than conservation groups to be run by volunteers alone, without any paid employees. Part of this difference is due to resource limitations; conservation groups have more members and larger budgets, which enable them to employ more people. Even taking these resource differences into account, however, multivariate analyses (not shown) find that ecology groups are still more likely to draw volunteers into the organization.

Several implications follow from the small staffs of most groups. First, the smallest organizations are often sustained by the efforts of a single individual. With a small staff, or no staff at all, the director determines the goals and political style of the group. He or she becomes the press representative, the primary strategist, and the major administrative officer. The success or failure of these small organizations rides on the entrepreneurial skills of their director. Second, small staffs mean less professional expertise within the organization. A mixture of professional skill and expertise is needed to compete effectively in contemporary politics. Few environmental organizations can afford to employ the lawyers, scientists, journalists, public relations experts, and other professionals that they might require in government hearings or policy-making meetings. Instead, the movement must depend on a network of advisers that is spread across many organizations (see also Lowe and Goyder 1983, 49–50; Berry 1977, 204–05). Within a nation, for instance, one group might have recognized specialists in the technology of nuclear plants, another in waste management, and so forth; such attributions of specialization were not uncommon in our discussions with group representatives. The effectiveness of policy is thus linked to a group's ability to develop and utilize a network of shared professional expertise.

Organizational Structure

The literature on social movements maintains that the internal structure of SMOs provides a key to understanding the behavior and the social implications of a movement. Paulo Donati claims that "the way in which social movements—or rather social movement areas—are internally structured is the crucial factor in understanding their action in society and the social system"

(1984, 838). Donati's views are substantiated by a large body of scholarship claiming that organizational characteristics predict the ability of a group to mobilize resources, engage in political competition, and attain its objectives.

The Weber-Michels theory of institutionalization—the iron law of oligarchy—is considered the starting point for most research on organizational structure (Weber 1978; Michels 1954). There are two lines to this argument. First, the theory contends that even if a social movement begins as an amorphous and fluid collection of individuals committed to a specific goal, once organizational structures are established, they have a tendency to become highly centralized and institutionalized. Second, centralization and bureaucratization can isolate a movement organization from its supporters and lead to the substitution of organizational maintenance goals for the movement's original social objectives. Max Weber saw these pressures as emanating from the gradual rationalization and bureaucratization that occurs as organizations establish themselves. Michels stressed the inequalities between elites and group members that leads to oligarchic tendencies within voluntary associations.

RM theory expanded on the Weber-Michels model by noting that a variety of factors, not just the Weber-Michels imperative, guided the structural development of SMOs (Zald and Ash 1966). At the same time, the organizational emphasis of RM led these theorists to agree with the first proposition—that centralized, formally structured movement organizations are more typical of modern social movements and more effective at mobilizing resources and conducting sustained political campaigns (e.g., Gamson 1975; McAdam 1982; Tilly 1978; Mirowski and Ross 1981). The professionalized and centralized group is presumably more effective in mounting a direct-mail campaign, in organizing a membership drive, in targeting its political activities on a specific goal, and in performing the other operational functions of a movement. Moreover, external pressures reinforce these resource considerations. The mass media prefer a single, well-known public personality to project the public image of a group; politicians have similar difficulty dealing with amorphous organizations that lack an authoritative spokesperson, and bureaucrats even more so. In short, RM theory agreed with the Weber-Michels thesis: a professionalized and centralized organization is more likely to survive and prosper in the political system of modern industrial societies.[16]

Resource mobilization theory also brought attention to the possible displacement of the original goals of social change through professionalization. Robert Salisbury (1969) emphasized the individual motivations of the entrepreneur as critical to an organization's success (also Gamson 1975). Furthermore, some criticism of RM theory has focused on the process of goal displacement if groups follow the path of centralization and institutionalization (e.g., Piven and Cloward 1979; Gerlach and Hine 1970).

One of the most dramatic claims of European research on NSMs, including research on the environmental movement, is that these movements violate the Weber-Michels imperative (e.g., Brand et al. 1986; Rammstedt 1978). Environmental groups, women's groups, and peace movement organizations often avoid professionalized, centralized, and bureaucratic organizations in favor of a fluid organizational framework based on a horizontal coordination of grass-roots groups in loose networks. In line with our model of ISA, analysts cite the ideology of these social movements as a primary reason for this alternative organizational structure. The New Environmental Paradigm, for instance, calls for the democratization of all areas of life through the expansion of citizen participation. The internal structure of ecologist groups, at least, should therefore reflect this orientation and serve as a measure by which potential supporters can judge the organization. As Joyce Mushaben argues in her description of European peace groups, NSMs are "committed simultaneously to the internal application of grassroots-democratic principles and the democratic transformation of external political institutions" (Mushaben 1989, 275).

Beyond ideology, NSMs might favor a loose organizational structure because of their diffuse and changeable social base (See Kuechler 1984; Donati 1984; Klandermans et al. 1988). Adherents of environmental groups, for example, are not integrated into a hierarchically defined social network that is easily mobilized. The type of closed social milieu—typical of the working-class movement (or comparable agrarian or religious networks), including labor union membership and an insulated social milieu created by a socialist party and its ancillary organizations—is simply lacking in the case of NSMs. A distinct cultural milieu does exist to support the activities of the green movement—countercultural networks exist in most urban areas— but these networks are created by individuals and not controlled or directed by any organization. Indeed, the antiestablishment values of ecologists are antithetical to the exclusive, cohesive, clientelistic associations that provide the basis of old social movements (Inglehart 1990; Cotgrove 1982). Thus, the style of elite-directed mobilization that characterizes old social movements is replaced by the fluid structure and persuasive style of ecology groups and other NSMs.

Evidence supporting the alternative organizational structure of NSMs comes from many sources. Locally based citizen action groups throughout Europe frequently display an aversion to hierarchy and formalized rules

(Guggenberger and Kempf 1984; Gundelach 1984; Donati 1984; Kriesi and van Praag 1987). The neighborhood locale and small size of these groups make an extensive organizational structure less necessary and less desirable. In addition, these tendencies apparently carry over to larger public interest organizations that emerge from these citizen movements. The coordinating committee of the West German peace movement and the Greenham Common peace protest in the early 1980s, for example, consciously rejected a centralized organizing structure as an extension of the ideology of the movement (Mushaben 1989; Rochon 1988). Similarly, many activists in the women's movement are openly skeptical of an institutionalized women's lobby that might come to dominate the movement; in the place of such an organizational structure, they prefer local action centers and self-help groups (Freeman 1975; Ferree 1987; Gelb 1989).

Apparently a decentralized and participatory organizational pattern also exists within the European environmental movement, especially among organizations with an ecological orientation. Antioligarchic tendencies are clearly visible among many ad hoc citizen action groups, where a small size and localized membership facilitate participatory politics (e.g., Andritzky and Wahl-Terlinden 1978; Guggenberger and Kempf 1984). At the national level of politics, most green parties in Western Europe display an aversion to hierarchy and bureaucratic procedures (Kitschelt 1989; Kitschelt and Hellemans 1990; Poguntke 1987, 1993). Participatory politics is central to the ideology of these parties and consequently becomes an active element in the green political identity. Similar tendencies apparently exist within the environmental lobby: analysts cite the antinuclear movement in many nations for its aversion to centralization and bureaucratization (Brand et al. 1986, chap. 2; Nelkin and Pollack 1981, 195–98; Rüdig 1991), and several ecology groups within the environmental movement also stress their participatory orientation.

This section considers a number of issues that evolve from this debate over the organizational structure of New Social Movements. We first examine the internal organization of environmental groups. Beyond formal democratic structures, we consider the extent of conflict within environmental groups, and the procedures that exist for conflict resolution.

Violating the Iron Law?

Our study of European environmental groups provides a partial test of whether the environmental movement is resisting the Weber-Michels imperative. We surveyed only organizations that have a continuing role in national politics on environmental issues. This requires at least a modest level of institutional development and continuity, including the ability to maintain membership records, compile relevant information, communicate with the membership, and perform the other functions of an ongoing political lobby. Our focus on national organizations also means that most groups have a geographically dispersed membership, which limits the ability of individuals to attend group meetings and contact group officials on a personal, face-to-face basis. Local citizen action groups or ad hoc national campaigns, where one would be most likely to find fluid structures and direct membership involvement, are excluded from this study by definition. Furthermore, groups vary in their ideological commitment to an alternative organizational structure; the stress on decentralization and participation is most closely identified with ecology groups, while conservation groups show little abhorrence to hierarchy and structure. In short, there are many practical reasons to expect that most environmental groups will follow a path toward professionalization and centralization; an ecologist ideology seems to be the only countervailing force against these tendencies.

During our interviews, one set of questions dealt with the opportunities for membership participation—whether there were annual meetings to elect officers and solicit opinions and whether the selection of local chapter representatives was democratic—and the significance of such input mechanisms for decision-making processes. A second set involved the decision-making processes within the organization, focusing especially on the role of group officials. We discussed the procedures for selecting the board of directors (whether it be the *Vorstand, conseil,* or *consiglio*) and the processes for decision making within this body. Other questions concerned the responsibilities of the organization's chief executive and relations between the executive and the board of directors. We often supplemented the interviews with information from the published rules of procedure or other official documents. Based on these materials, the interviewers coded several summary measures of the role of the membership and elites in guiding each group. Although this material is not as rich and extensive as a case study approach to organizational behavior, it does provide an overview of the organizational tendencies among environmental interest groups.

Our most general measure tries to summarize the overall organizational tendencies of mass-membership groups, focusing on the formal structures of the organization (table 4.6). Are groups structured to concentrate authority and responsibility at the center, or do they follow a decentralized structure? This overall assessment finds that most groups (81 percent) are built around

TABLE 4.6

Locus of Decision Making

	Conservation	Ecology	Total
Group Structure			
Centralized	91	63	81
Locally based	9	21	13
Membership-based	0	16	6
Total	100%	100%	100%
(*N*)	(23)	(19)	(48)
Possibility for Member Input			
Donate, no formal input	14	35	20
Weak opportunities	77	29	60
Significant opportunities	9	35	20
Total	100%	99%	100%
(*N*)	(22)	(17)	(45)
Major Decision Makers			
Chief executive officer	35	22	28
Board of directors	65	39	57
Staff	0	17	6
Local groups or members	0	22	9
Total	100%	100%	100%
(*N*)	(23)	(18)	(47)

Note: Figures are based on all mass-membership groups for which sufficient data were available.

a centralized structure, and only a small percentage place the locus of power with their membership.

To add more depth to this general assessment, we determined separately whether mass-membership groups have mechanisms for direct involvement by members, without considering whether these procedures are effective as a vehicle for participatory decision making. For instance, more than three-fourths of mass-membership groups elect their officers, usually at an annual general meeting (AGM) attended by members or representatives of local groups.[17] Yet this formal electoral procedure provides only a minimal level of membership participation, because as a rule the elections at general meetings simply ratify selections already made by the leadership of the organization. Many of our respondents readily acknowledge the de facto control of nominating committees in identifying and screening prospective candidates. In addition, the proportion of members who participate in these elections is extremely low. Seldom does participation in an AGM exceed more than 10 percent of the membership.

Not only is membership participation limited, but group representatives feel that members do not want to participate, even though most officials would welcome such involvement. This is not surprising, because it illustrates Olson's (1965) theory of collective action. Writing a check for annual membership is not time-consuming and the fees are low, but attendance at an AGM makes substantially greater demands upon an individual. (It is also easier for the average member to voice dissent by exiting one environmental group and joining another than by working to reform a group from within.) Simply put, most voluntary environmental associations are not vehicles for extensive citizen participation, and most group representatives openly acknowledge this fact.

In examining the influence of membership on decision making within mass-membership groups, we found that in 20 percent of the groups the members have no formal role at all—they exist simply to contribute money to the organization and have no other formally recognized status (table 4.6). In fact, these groups sometimes refer to members as "donors" or "contributors" to denote their status. About two-thirds of the groups have some procedures for membership input, but because of the factors discussed above, these procedures are largely ineffectual in influencing group decision making. Significant membership influence exists only in a minority of environmental interest groups, and in most cases this input comes only indirectly through locally elected membership representatives.

We also found that the procedures for decision making within environmental groups convey an image of power concentrated at the top of the organizational structure (lower panel of table 4.6). Most groups try to be responsive to the general interests of their members, but more than 80 percent of the mass-membership groups we surveyed nonetheless follow a fairly hierarchic (or oligarchic) structure for making major decisions.[18] In most instances (57 percent), the board of directors is primarily responsible for determining policy and administering the organization. Reflecting our earlier discussion of the importance of an individual entrepreneur, the chief executive officer exercises personal direction and control of the organization in 28 percent of the groups we surveyed. Fewer than 10 percent of the groups function on a more democratic model of group decision making, and in most of these cases participation is based on representative democracy through participation by delegates of local groups.

Many environmental interest groups see themselves as supplying a product, which individuals can support (or not) through their membership in the organization. Such "marketing decisions" generally come from top officials rather than percolating up from the membership. One respondent frankly stated that the group leaders had undertaken a campaign to protect tropical rain forests because of the importance to the environment, without knowing how the membership would respond to a campaign that differed markedly from past actions. Other respondents reported that staff members selected campaigns that they thought would generate more popular support for the organization, while concealing from members many lobbying activities and administrative actions. Our interviews are replete with such examples of decision making from the top down, and most group representatives believe in the Weber-Michels imperative, even if this is unstated.

The oligarchy and centralization in European environmental groups mirror a pattern found for most British environmental interest groups (Lowe and Goyder 1983, 51–55) and American voluntary associations (Schlozman and Tierney 1986; Berry 1977). Most environmental interest groups are apparently not immune to the Weber-Michels imperative.[19] Despite their advocacy of the public interest, the maintenance of the organization and a marginally involved membership lead to centralization of authority and administration in most groups. Throughout the discussion of organizational structure and decision making, group officials offered numerous illustrations of the need for effective leadership and centralized administration. Organizations need authoritative decision makers, and such authority nurtures oligarchy. Similarly, the simple desire for efficiency that motivated one national group to consolidate the membership records of its county associations into a single centralized computer file (also giving the national organization greater control over communications and fund raising) illustrates a logic that affects most groups.

In spite of these general tendencies among environmental groups, some mass-membership groups are able to avoid these organizational tendencies in favor of more participatory and democratic structures—these choices may be influenced by environmental orientations.[20] Our framework of ISA suggests that the political values of ecology groups will lead them to adopt participatory structures within their organizations; these groups supposedly epitomize the NSM model. Because conservation groups lack these values, or at least give them less importance, they are more likely to conform to the organizational tendencies expressed by RM theory.

The first panel of table 4.6 indicates that ecology groups are more likely than conservation groups to provide a formal structure that is based on local groups or the membership itself (tau-c = .21). Furthermore, our coding of the possibility for membership input finds that opportunities are more common for ecology groups. For instance, the FoE groups in our study are generally noted for their participatory style, but in most nations this is based on consultation with local group delegates. Only the smallest FoE groups are able to involve individual members directly in the decision-making process. The Danish ecology group NOAH illustrates one of the most open organizational structures we encountered (Jamison et al. 1990, chap. 3). NOAH is an association of local citizen groups throughout Denmark who turn to a central office for financial and political support. Representatives of local groups meet twice a year to discuss issues, set campaign goals, and define the agenda for the organization. In addition to these semiannual meetings, local groups can suggest new projects or request support from NOAH by circulating the proposal in the group's newsletter, which then serves as a forum for communication among the groups.[21] NOAH's lack of paid employees facilitates this lack of oligarchy.

The third panel of table 4.6 indicates that most ecology groups are controlled by their leadership, although these groups are also more likely than conservation groups to adopt democratic, decentralist structures (.29). In nearly a quarter of the ecology groups, members or representatives of local groups make the most important decisions for the organization; this occurs in none of the conservation groups we surveyed!

Such participatory norms obviously do not permeate all ecology groups; Greenpeace is the most notable exception to this pattern. Greenpeace views itself as an elitist group directed by its staff and those who actively participate in its campaigns (Eyerman and Jamison 1989). In most nations, individual contributors to Greenpeace are merely financial sponsors and lack any formal standing as members. At one time, this image has created some disdain for the group among the public. In their study of British environmental groups, Lowe and Goyder (1983) were told by a Greenpeace official that individuals should contribute money but then leave the organization alone; a Greenpeace representative in our survey stated that their members "could only vote with their feet." Five of the six Greenpeace organizations surveyed report having no input structure for members. As a result, environmental ideology is not correlated with opportunities for membership input. If we exclude the Greenpeace groups from the sample—because they obviously do not share this aspect of ecological orientation—the correlation among the remaining groups rises to .30. Furthermore, even Greenpeace groups realize the tension between their antiestablishment rhetoric and their oligarchic structure. In our interview, the chairperson of Greenpeace-Britain stated that Greenpeace needed to involve its members in the organization, "otherwise (Greenpeace) would remain just a political gadfly." At the same time, however, he noted that this reform would stop if the staff felt it diverted their energies from action campaigns (a pattern they attribute to FoE).

Ecologists are drawn toward new ways of thinking, but they are also sensitive to the tensions between organizational effectiveness and participatory decision making. Whereas some ecologists may resist the iron law and some groups (such as FoE and NOAH) may even use their participatory style as part of their political image, they are not immune to its forces.

Conflict Resolution

The issues of oligarchy and decision-making structures are often linked to the extent of internal consensus within an organization. Doug McAdam (1983) claims that participatory democracy can function in small organizations and those with high consensus; but when there is division within a group, the participatory approach lacks the means to limit the length and intensity of conflict. The research of Stephan Barkan (1979) and Gary Downey (1986) on American environmental groups seemingly substantiates this position. They find that political and ideological disagreements in groups opposing nuclear power can immobilize the organization, as endless discussions and consensus-building efforts turn decision making into a war of attrition.

From the RM perspective, the organizational patterns existing among environmental interest groups may be, at least in part, related to the amount of internal dissensus among supporters. We have repeatedly stressed the diversity of environmental ideologies and identities that exist within the movement. This diversity would suggest that organizations prominent enough to be included in our study would have adopted centralized decision-making structures as a by-product of their growth and institutionalization; those who do not follow this pattern would be subject to greater intraorganizational conflict and might not endure.

On the surface, our findings seem to support this explanation (fig. 4.5). We found that mass-membership groups with participatory decision-making

FIGURE 4.5

Environmental Orientations and Organizational Characteristics   Figures are tau-c correlations based on 47 mass-membership groups; the statistics in parentheses are Pearson correlations. All relationships are statistically significant (p < .10).

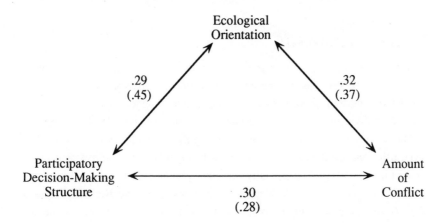

structures experience greater internal conflict.[22] Sometimes this conflict occurs among leaders as they debate goals or tactics; at other times, tensions arise between the leadership and members.

On closer inspection, however, this relationship between organizational structure and internal conflict may be a spurious effect of the environmental orientation of a group (fig. 4.5).[23] Ecology groups tend to have participatory decision-making structures. In addition, ecologists are, by definition, challenging prevailing social norms, and this orientation is more likely to generate disagreement over tactics or goals within the organization. By comparison, conservation groups are both more hierarchically structured and more concerned with consensual goals that hold less potential for conflict. For instance, many groups acknowledged that their members are divided on the issue of nuclear power. Conservation groups generally avoid this division by defining the issue as outside the purview of the organization. Ecology groups, however, are often forced by their members or leaders to take a stand; after delineating their position they attempt to resolve the internal differences on the issue.

Our findings thus suggest that the organizational tendencies of ecology groups are distinct from those of conservation groups. The environmental orientation of ecology groups encourages them to develop participatory structures *and* to deal with conflictual issues. Debate and disagreement is part of the ethos. We can conclude from this that intragroup conflict is primarily a function not of the structure of an organization, as some RM mobilization theorists maintain, but of the ideological content of the movement.

A Pattern of Diversity

The literature on social movements tends to adopt a dichotomous view of SMOs. On the one hand, classic organizational behavior theory (Weber-Michels) and RM theory project a logic that "bigger is better." Large, centralized organizations, they argue, are more efficient in mobilizing the resources of a social movement and utilizing them effectively (Gamson 1975; McCarthy and Zald 1977). On the other hand, other social movement theorists and some analysts of socially marginal groups echo E. F. Schumacher's (1973) motto that "small is beautiful." Frances Piven and Richard Cloward (1979) maintain that institutionalization is deadly to a social movement because it ends spontaneous growth and leads a group to work within the established social order. Similarly, Luther Gerlach and Virginia Hine (1970) claim that decentralized movements characterized by a division of labor and integrated by informal networks (what they call *SPIN* networks) are more effective than highly organized groups.

Our investigation of the structural characteristics of environmental groups indicates that the movement contains a diversity of organizational structures: groups may be large and small, centralized or decentralized, hierarchic or participatory. Most important, environmental groups represent a variety of political viewpoints. Green politics includes a wide array of issue interests, spanning concerns about nature preservation and cultural heritage to the technologically complex problems of nuclear power and global warming.

We believe that the prime cause of this political and organizational variation is the diversity of views and values within the environmental movement itself. Some individuals are motivated by issues of wildlife conservation and preservation of the natural environment, others by problems of acid rain and toxic wastes. Some people are inspired by the exploits of campaigning groups, whereas others favor more policy-oriented activities. Moreover, some see environmentalism as reflecting a larger vision of social change; others see only the need for more bird sanctuaries or better enforcement of existing regulations. Given this diversity in the political content of environmentalism, mass-membership groups attempt to differentiate their "product" and establish their place in the environmental movement industry.[24]

One might rightly claim that each of the groups we surveyed has a distinct identity—a necessity if a public interest group is to endure and prosper. Just like consumer products competing in the economic marketplace, environmental groups are competing for support and influence in the political marketplace. Once conservationists establish the RSPB, there is no need to duplicate its goals and methods in a second organization. Indeed, a second organization would likely fail unless it can improve on the RSPB model. Instead, new firms (or public interest groups) enter the market by offering new products or innovation on past offerings. Our interviews included numerous illustrations of such differentiation.[25]

The leadership of SMOs is another source of their diversity. The personal values and "marketplace" estimates of environmental elites can lead to the creation of new groups and alter the behavior patterns of existing groups. Indeed, environmental groups have normally begun through the actions of a few individuals; their founders' decisions about the goals and nature of a group created an identity that determined the initial success or failure of their efforts. In many of the smaller groups the impact of the director is visible throughout the organization, making the group's identity an extension of his or her personal political beliefs.

Political conditions also can influence the structure of environmental groups and the organizational diversity within the movement. For instance, government regulations, especially provisions of the tax law, can proscribe the legal structure of voluntary associations. Such regulations may also nurture the variety of organizational structures; Greenpeace UK has both a tax-exempt foundation and an independent political arm. In other cases, governments might provide support for educational or research activities, which lead to the creation of educational foundations and research institutes.

Finally, the diversity of environmental organizations represents the different functional needs that exist within any social movement (Zald and Ash 1966). Mass-member groups exist to mobilize public support for the movement and to provide political representation for citizens interested in environmental issues. National umbrella groups work to coordinate the activities of individual groups, or at least provide a framework for formal and informal interaction. Research institutes and educational organizations create and disseminate knowledge on environmental issues.

Chapter Five

ENVIRONMENTAL ELITES

How European environmentalists perceive their organizations and the larger political world provides essential information for understanding the motivations and actions of these groups—in part because of the decisive role of leaders in policy making. We also found that in many cases the organization *is* the individual and that one of the organization's most precious resources is the skill and ability of its leadership.

The personal characteristics and political beliefs of group officials furnish valuable information on the nature and underlying political tendencies of the environmental movement. Who are Europe's environmental activists, what paths did they follow to reach their present positions, and what are their personal political values? There is abundant speculation about the social characteristics of environmental elites that has implications for how the movement should be interpreted in larger social and political terms.

Even more intense debates focus on the political values and beliefs of environmental activists—who have been linked to virtually every political orientation imaginable. Many critics look at the green movement and see red; environmentalism is viewed as a vehicle for radical leftists advocating revolutionary political and economic change. Other analysts paint environmentalists in brownish tones; descriptions of the German Greens as a revival of the Hitler youth movement have appeared in both the popular press and social science journals. To some they are "left-libertarians," to others reactionaries. Numerous political labels have been applied to the environmental movement (see the historical review in Bramwell 1989). Clearly, knowledge about how environmentalists think about the world of politics, how they view environmental and other political cleavages, and where they fit within this world can illuminate the political forces that are leading the green movement. Behind what the movement says about itself, what do its leaders really think about the political controversies of advanced industrial societies and the process of democratic governance?

Social Background

It is probably inevitable that the leadership of most political groups is socially unrepresentative of the public at large. Senior officials in nearly any political party, major national interest group, or political association are more middle-class and better educated than the general population; such social traits are routinely preparation for, or even a prerequisite of, elite positions in society or politics (Putnam 1976, 33). But in the case of environmental groups and other NSMs, the social composition of the leadership is intertwined with questions about the political orientations of the movement.

One of the most frequent criticisms of environmentalism is that it is an elitist movement and that its political views represent those of middle-class privilege. Environmentalism is sometimes portrayed as a leisure activity of an affluent class that is out of touch with the true needs and wants of the average citizen, and therefore unrepresentative of society as a whole. William Tucker has stridently commented that environmentalists are "people who have benefitted from the economic system much more than the average person. Then instead of wanting or allowing others to do the same, they have set their sights much higher than normal. Their values and positions are those of a nation's aristocracy" (1982a, 43). In more thoughtful and cautious tones, other scholars at least partially link the recent explosion of citizen action groups and single-issue associations to the expansion of skilled (and underutilized) entrepreneurs created by the rapid growth in university enrollments (Alber 1989). Jeffrey Berry (1977, chap. 4) has similarly documented the middle-class base of public interest groups in America. Environmentalism is generally seen as a movement of the middle class.

Certainly the leadership of the environmental movement has always had a distinctly middle-class accent. This has been especially pronounced in nature conservation groups. For instance, when the RSPB received its royal charter in 1904, it had three duchesses, three earls, a marquess, a viscount, and three bishops among its vice-presidents—hardly the makings of a revolutionary movement. Most of the other conservation groups formed at the beginning of this century were similarly drawn from elitist circles: the FFPS was founded by a former British colonial governor of Uganda, the Association for the Preservation of Natural Monuments drew upon Amsterdam's elite, the SNPN was formed by leading French zoologists and naturalists, and DN was established by members of Denmark's social and cultural elite. This pattern carried over to the formation of conservation groups in the 1960s and 1970s. Among these, the establishment link is perhaps best illustrated by the WWF. Early leaders included members of royal families (Prince Philip headed WWF-Britain and Prince Bernhard of the Netherlands was the president of WWF International) and Europe's social elite (Mrs. Giovanni Agnelli was the first president of WWF in Italy).

Ecology groups also appear to draw their activists and leadership from the middle class, but without the elitist overtones of the conservation movement. Ecologists are portrayed as "middle-class radicals," representing those who are alienated from the prevailing social order or who participate only marginally in that system. Wilhelm Bürklin (1987b) partially attributes the formation of the German Green Party to the frustrations felt by unemployed young or underplaced academics, who responded by creating their own organizations to challenge a system that was ignoring their needs; Jens Alber (1989) endorses this interpretation. Claus Offe (1985) describes NSM activists as partially drawn from the "de-commodified" sectors of society; this includes the unemployed and the economically marginal elements of the middle class. Stephen Cotgrove (1982) similarly finds that British environmentalists are disproportionately drawn from nonbusiness sectors of the middle class: teachers, government employees, and the creative professions.[1]

These discussions of the socioeconomic basis of environmental activism incorporate two assumptions. First, activists are disproportionately drawn from upper-class groups with the resources (education, organizational skills, and income) to allow them to be politically active. Second, within the middle class, environmentalism is a movement challenging the prevailing sociopolitical order and should therefore attract individuals from professional strata who are likely to question the status quo.

The discussion of the leadership of NSMs has also focused on gender as an important criterion for elite selection (Steger and Witt 1989; Stern et al. 1993). The political alliances and personnel ties between ecologists and the feminist movement are readily apparent in most nations; indeed, such bonds contribute to their general classification as NSMs. This overlap is partially explained by practical political realities: in challenging the political status quo—though in different areas—each movement relies on the other for support, because alliances with the political establishment are limited. It is also claimed that the values of both movements spring from a common source. Women are socialized into nurturing roles that lead them to be more sympathetic to environmental issues and the New Environmental Paradigm. The feminist movement contributes to the cultural transformation that ecologists seek, in forms described as eco-feminism or feminist environmentalism (Merchant 1980; Warren 1987; Diamond and Orenstein 1990). As a consequence of the political alliances and shared value structure of these two movements, we would expect women to find easier access to leadership positions within the green movement, more so than among traditional economic or political groups. Indeed, progressive elements of the environmental movement, especially green parties, are vocal advocates of women's rights and aggressively recruit women to positions of leadership. Can these patterns be generalized to the European environmental movement as a whole?

If these presumptions about environmental elites could be summarized, the claim would be that environmentalists, or at least ecologists, represent a so-called counter-elite to the existing political establishment. They are supposedly recruited from among the young, highly educated intellectuals of contemporary societies, providing positions of influence for those who question the dominant values of advanced industrial societies.

We might attempt to answer the following questions concerning social composition in absolute terms: what percentage of group officials are drawn from the middle class, and what percentage are female? But because virtually all politically elite groups are unrepresentative of the general public, the question should be posed in relative terms. Are environmentalists more middle-class, younger, or more female than other elite groups? Therefore, we compare environmentalists to a sample of European political elites who were candidates in the 1979 elections to the European Community Parliament (Inglehart et al. 1980). These candidates for the European Parliament (known as CEPs) represent high-level political elites, though they are perhaps not as exclusive as members of the respective national parliaments. In general terms, the political stature of CEPs seems roughly comparable to being the head of a national environmental group. The CEP sample also has the undeniable advantage of being the only broad European sample of political elites that has been conducted in recent years and that includes all the nations in our study except Greece. In addition, the CEP study includes several questions that were replicated in our survey, providing yet another comparison.

In terms of educational level, environmentalists are as well educated as candidates for the European Parliament (table 5.1). Four-fifths of both elite groups have attended university, whereas the comparable figure for the European public is closer to one-fifth. What is socially distinctive about environmentalists is that their class backgrounds are so narrowly focused. Over two-thirds of environmentalists come from professional occupations (another

TABLE 5.1

Profile of Environmental Elites and Europarliament Candidates

	Environmental Elites	Europarliament Candidates
Average age	41 yrs.	43 yrs.
Female Representatives	18%	18%
Education level		
Secondary or less	12%	20%
University	55	26
Graduate degree	33	55
Total	100%	101%
Occupation		
Scientist	18% ⎫	⎫
Government employee	12 ⎪	⎪
Journalist	8 ⎬ 68%	⎬ 35%
Teacher	14 ⎪	⎪
Other professional	16 ⎭	⎭
Businessperson	8	32
Manual worker	0	21
Other	2	10
Student	21	2
Total	99%	100%

Source: Data on Eurocandidates are from the 1979 Study of Candidates to the European Parliament; Inglehart et al. (1980).

one-fifth went straight from the university to working for the group). Teachers, natural scientists, journalists, and public administrators are heavily overrepresented within the environmental movement—this is a movement led by the new middle class. Not one of our respondents had been a manual worker before joining their group, and only 8 percent had been employed in business. Political elites are a more socially heterogeneous group, drawing significantly upon the working class (21 percent) and business (32 percent) for their membership. Furthermore, ecologists generally replicate the overall pattern of social characteristics for environmentalists.[2]

The profile of environmentalists does not differ markedly from European political elites on other social characteristics. Despite the rhetoric of some green activists, women are as poorly represented within environmental interest groups (18 percent) as they are among political elites. Even among ecology groups, women constitute only 24 percent of our respondents. Furthermore, our female respondents were more likely to be found at lower levels of the organizational hierarchy and in small underfunded groups. Environmentalists are slightly younger than European Parliament candidates, but these differences are relatively modest considering that the EC Parliament is often described as a retreat for elderly politicians.

In summary, the leadership of environmental groups has obvious origins in a distinctive stratum of the middle class. Environmental elites disproportionately come from professions that specialize in the creation and application of information and symbols: scientists, teachers, journalists, and public administrators. These careers seem to combine both the political orientations that lead to environmental activism and the entrepreneurial skills that are required by public interest groups. How does one interpret these findings? We would stop short of treating such data as evidence supporting the middle-class radicalism thesis (cf. Kitschelt and Hellemans 1990, 103–05). Environmental elites are predominately male, middle-aged, and middle-class—characteristics not associated with radicalism. As we shall see, the radicalism of environmental elites is better judged by their political values than their social backgrounds. In addition, we would also question the argument that the environmental movement represents an important alternative career ladder for counterelites who are excluded from normal career pathways. Many of our respondents have had successful careers in their chosen profession and could be more financially successful if they returned to those careers. Instead they choose to work within the environmental movement because it represents more than just a career to them. The low representation of women, even among ecology groups, further weakens the argument that environmentalism provides an alternative career ladder for counterelites. Even among ecology groups, the social profile of their leadership is more representative of middle-class volunteerism than middle-class radicalism.

Environmental Careers

Our discussions with group officials made it apparent that many of them define their careers in terms of the movement. The growing importance of environmental issues has created an employment opportunity for professionals with a public service orientation. Environmentalism is a social movement

industry that provides opportunities for employment in the public and private sectors, and the varied career backgrounds of our respondents illustrates this point (table 5.1). People circulate within the movement: natural scientists become active in conservation groups, activists enter the government to work on environmental policy, educators combine their academic and political interests, consultants work for the public and private sector, and unpaid volunteers from one green group find paid employment in a better-funded group. The extent of such career opportunities has important implications for political recruitment into the movement and the effective functioning of groups within the contemporary political process. For instance, a paid professional staff is likely to attract a more skilled and experienced leadership and to mount a more serious policy challenge than are volunteers and political novices.

We surveyed the top officials of established national interest groups and found, not surprisingly, that about two-thirds of our respondents are employed by the group they represent. The remaining one-third work on a voluntary basis and consider environmentalism an avocation. Still, many of the volunteers might be considered professional environmentalists because they head a small organization that they founded or maintain.

The work experience of our respondents clearly reflects the professionalization of the environmental movement. On average, they had worked more than eight years for their present environmental organization; on a percentage basis, this translated into more than half of their professional lives.[3] In addition, more than one-third had previously worked for another environmental group. If one were to include environmentally related work for other public and private organizations, then it is clear that most of these environmentalists are treating the movement as a profession and not merely as an avocation.

The professionalization of this movement is common to other public interest groups (Schlozman and Tierney 1986). Our study focused on only one element of this career path, circulation within environmental movement organizations. To tap into personnel networks and informal ties existing within the movement, we asked whether respondents had worked for another environmental group, and if so, which one. We were especially interested in the personnel links between conservation and ecology groups that might provide informal bonds between these two components of the environmental movement.

In studying employment circulation we found that ecology groups provide an entry point into environmental activism, undoubtedly because of their youthful and ideological orientations and their reliance on young volunteers

FIGURE 5.1

Employment Circulation among Environmental Groups   Figures are the percentages
of respondents within each group (at the head of each arrow) who had prior job
experience with another group; none indicates those whose prior work experience
had not included an environmental group or whose first job was with an
environmental group.

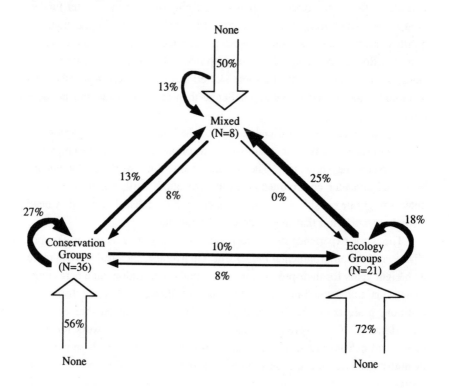

(fig. 5.1). Within the ecology groups we studied, 72 percent of the individuals
had never worked for another environmental group, and 18 percent had
previously worked for another ecology group. This might occur because of
the strong ideological pull of these groups for younger Europeans. In addition, work for a small ecology group provides career training in environmental politics, which can lead to better-paying and more secure jobs in
government, research, or more established conservation organizations. In
fact, 25 percent of the representatives of mixed environmental groups re-

ported that they had previously worked in an ecology group, as had 8 percent of the respondents from conservation groups.

If ecology groups are an entry point into green politics, conservation groups generally represent a step up the career ladder, because most of these groups have a larger resource base and can offer more secure job prospects. Nearly half of the representatives in conservation groups had previously worked for another environmental group, and the flow between conservation groups and other environmental groups appears to be almost evenly balanced. For instance, 8 percent of the officials in conservation groups had previously worked for an ecology group, while the reverse flow was 10 percent. In fact, these nearly equal percentages reflect an imbalance, because the professional staff of conservation groups is two to three times larger than the combined staffs of ecology and mixed environmental groups. For the percentage flows to become equal, a larger absolute number of people would have to be moving into conservation groups from other parts of the movement.[4]

Political Values and the Democratic Process

Nearly everyone agrees that environmentalism represents an important new political force in advanced industrial societies; the disagreement lies in the nature of this political force. The writings of some green activists and the political activities of others have projected strong, and often contrasting, images of what environmentalists think about the political issues and political processes of Western democracies. One frequent characterization pictures environmentalists as surreptitious leftists (or worse) who are merely using the cloak of environmental politics to cover their true colors. Nicholas Ridley, the former British environmental secretary (1986–89), once labeled Britain's Green Party as a group of pseudo-Marxists. Helmut Kohl has a well-worn campaign phrase in which he describes environmentalists as tomatoes: they start out green and end up red. Giulio Andreotti used the analogy of a watermelon: green on the outside, red on the inside. Coexisting with these leftist images of environmentalists is a contrasting picture of activists as naive idealists feasting on bean sprouts, listening to whale songs, and holding alternative political views. One can also point to examples of eco-socialists, eco-feminists, red-greens, brown-greens, and green-greens. In short, hypotheses about the political roots of the environmental movement appear limitless (Bramwell 1989; Scott 1990).

Our survey probed the personal political beliefs of group representatives as a means of tapping the underlying political values that drive the green movement (or at least its leadership). The actual political beliefs of our respondents can explain how the diverse ideological and political themes of environmentalism are reconciled by the individuals who hold positions of influence and authority. In other words, what is the mix of political values represented among the leaders of Europe's environmental groups? This information is related to the political origins and alliance options of the movement; for example, strong socialist views suggest a potential for closer ties to the traditional Left. In addition, given the importance of leadership in the direction of these groups, the views of our respondents provide a measure of the latent political tendencies that may influence the course of environmental action. Their views should be considered along with members' views and prior organizational behavior in forecasting the directions of environmental action.

We are, therefore, not primarily concerned with what our respondents think about environmental policy and reform. Rather, we consider how leading European environmentalists think about contemporary political issues and the political process and how they locate their views within the larger framework of current political debate. What, in a general sense, is the broad political identity of environmental activists?

Left, Right, or None of the Above?

Contemporary political discourse often relies on the terms "Left" and "Right" and on the Left/Right framework as an organizing concept for evaluating issues, politicians, and political parties. Many citizens use these terms to orient themselves to the world of politics, even if their understanding of these labels is not specific (Fuchs and Klingemann 1990; Inglehart 1990). For political elites, Left and Right signify a rich, and often diverse, set of political beliefs. In this sense, Left/Right self-image identifies an individual's overall political orientation and perceived position on current political issues.

Despite, or perhaps because of, its common currency, the Left/Right dimension has been a contentious symbol for environmentalists. Because the terms are predominately identified with the traditional socioeconomic cleavages of Western societies, many environmentalists reject the application of the Left/Right framework to the movement and their own political viewpoints. The slogan "Neither Left nor Right, we are out in front" has become a standard political disclaimer of the movement. Yet we have also noted that others are eager to label environmentalists in just such Left/Right terms.

With this background, we began our inquiry by asking environmentalists to locate themselves on the same Left/Right scale used in other European public and elite surveys (fig. 5.2). In spite of their rhetorical rejection of the

FIGURE 5.2
Left/Right Political Self-placement of Environmentalists and Environmental Groups

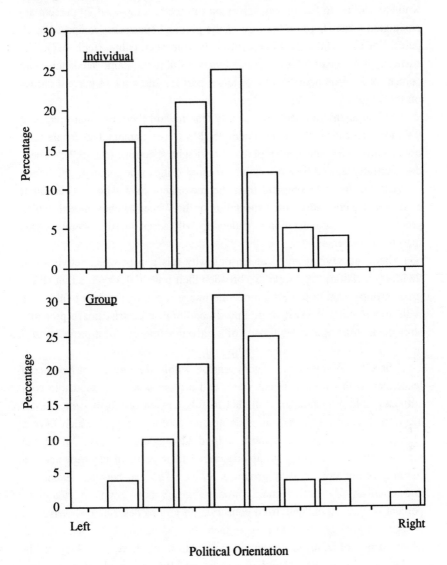

framework, the vast majority of environmentalists locate themselves on the Left/Right scale—and display an unmistakable tilt leftward.[5] Among those who are willing to locate themselves on the scale, a lopsided 79 percent are positioned somewhere on the Left of the political spectrum, compared to 69 percent of CEPs and only 43 percent of the European public (table 5.2). The partisan preferences of environmentalists exhibit the same tendencies: 46 percent of environmentalists favor leftist parties, and only 11 percent parties on the Right.

Although environmentalists clearly lean toward the Left, their images of the movement itself are more moderate. Most environmentalists locate their organization near the center of the Left/Right spectrum (fig. 5.2).[6] These data partially substantiate the claims of those critics who maintain that green activists are more ideological than the movement as a whole. This initial evidence also contradicts the expectation of the Weber-Michels model, which predicts that an organization's leadership will adopt more moderate views than the membership.

There are several practical reasons why this gap between activists and their organizations may occur. To broaden their potential for recruiting allies, green groups want to project a moderate image, opening the door to alliances with liberal and conservative groups. In addition, a centrist image probably strengthens the popular legitimacy of a citizen action group; it projects itself as independent of, and above, traditional political interests.

There is, in any case, a latent tension within the movement between a leadership with strong political views and a more politically moderate constituency. Many respondents alluded to this tension during the interviews. This ideological gap is not as wide among ecology groups, which have a clearer leftist image and a more unified ideological identity. All of the representatives of ecology groups described their personal ideology and the political orientation of their group as leftist.[7] The gap between leadership and organization occurs primarily among conservation groups. Although 72 percent of the respondents from conservation groups describe themselves as leftists, only 47 percent see their organization as having leftist tendencies.[8] Many senior officials in conservation organizations have a background in ecology groups or other New Left groups, and their political values are thus more liberal than their membership. The head of one conservation group, a former FoE activist, explained that the members of his group were more moderate in their environmental views than he was; but he also stated that one of his goals was to educate the membership on how broader ecological problems, such as industrial wastes and acid rain, harmed wildlife habitats and thus should also be a concern for them. This ideological gap should

ENVIRONMENTAL ORGANIZATIONS

TABLE 5.2

Political Beliefs of Environmental Elites, the European Public, and Europarliament Candidates

	Environmental Leaders	Europarliament Candidates	European Public
Political orientation			
Left (self-identification)	79%	69%	43%
Traditional economic issues			
Control multinational corporations	90	93	—
Reduce income inequality	85	93	79
Increase government role in economy	54	51	62
Expand public ownership of industry	21	33	33
New politics			
Postmaterial values	67	32	14
Stronger measures for sex equality	98	—	—
Freedom to have abortion	88	71	—
Limits to growth	85	—	—
More rights for employees	84	65	—
Storage of nuclear waste unsafe	82	—	—
Develop nuclear energy	12	72	54
Foreign policy/Peace			
Stronger defense effort by Europe	22	61	44
Ensure security by disarming both superpowers	94	—	—

Note: Figures represent the percentage of respondents agreeing with the policy. Sources: 1979 Study of Candidates to the European Parliament; public opinion data are from Eurobarometer 19 (April 1983). Dashes indicate that data were not available.

restrain the influence of leftist elites within conservation groups, while also providing a latent support for liberal positions within the conservation movement because of their more leftist leadership.

Although we have discussed environmentalists and their organizations in terms of broad Left/Right tendencies, it would be incorrect to label environmentalists simply as leftist in the traditional sense of the term. About one-sixth of our respondents refused to position themselves on the Left/Right scale (versus less than one-tenth of CEPs). In addition, other environmentalists said that they were uncomfortable with the traditional definitions of Left and Right based on class and economics. Several volunteered that there were different dimensions of Left and Right and that their beliefs were not easily conceptualized in terms of traditional political labels. Therefore, the meaning of Left and Right to environmentalists, and hence the interpretation of these political identities, cannot be assumed without looking more closely at political beliefs.

Old Politics and New Politics

In assessing the political beliefs of environmentalists, we chose issues that we thought would involve the political cleavages that structure competition in contemporary Western democracies. Like some of our respondents, we saw the Left/Right framework as encompassing two broad dimensions of conflict (Inglehart 1984; Dalton 1988, chap. 7).[9] First, there are the traditional issues of Left/Right division: socioeconomic conflicts and attitudes toward the economic role of the state. Second, there are issues of postmaterialism or New Politics, which supposedly define the political base of environmentalism and other NSMs, such as the women's movement and peace movement. Equally important, we tried to replicate questions available from other European surveys so that opinions could be judged against a reference standard.

Regardless of their leftist image and self-defined ideology, environmental leaders do not differ greatly from top-level European political elites (the Eurocandidate sample) or the European public on basic economic controversies (table 5.2). Most environmentalists seem satisfied with the present scope of economic activity by government; only a minority (21 percent) favor expanding public ownership of industry, and a bare majority (54 percent) prefer an increased government role in managing the economy. Our discussions often elicited open acknowledgments that the Old Left model of an interventionist state conflicts with the populist and decentralist values of the environmental movement. This is also an opinion that spans conservation and ecology groups. For instance, one Greenpeace official spontaneously noted that "we do not desire a traditional eastern bloc planned economy," while the heads of several conservation groups said nationalization in their country had already progressed far enough. In addition, a large majority of environmentalists favor actions to control multinationals or to narrow income inequality—two other Old Left issues—but these opinions are shared by European citizens and political elites.

Thus, the distinctive aspect of environmentalists' views on these traditional economic controversies is their moderation. Where comparable survey questions were available, environmental leaders were slightly more *conservative* than the public or political elites on the economic issues that have traditionally defined the meaning of being Left in European politics. For instance, although only one-fifth of environmentalists favor further nationalization of industry, one-third of the CEPs and European public endorse such a position. It seems therefore that the leadership of environmental interest groups is hardly deserving of a radical leftist image.

The liberal identity of environmentalists is more distinctly derived from their views on the issues of advanced industrial societies, the so-called New Politics. For instance, the clear majority (67 percent) give priority to post-material goals, such as freedom of speech and political participation.[10] In contrast, the average European citizen (and CEP) is still primarily concerned with material goals such as social order and inflation.

Questions tapping green themes predictably illustrate the strong opinions of environmentalists on these issues. Most group representatives (85 percent) reject the assumption of limitless economic growth that underlies the modern industrial state, a belief that Milbrath (1984) argues is a key component of an ecologist's commitment to a sustainable society. Opposition to nuclear power is one of the major policy initiatives of the green movement. Environmentalists almost uniformly (82 percent) feel that there is presently no safe way to store nuclear wastes. Similarly, few group representatives (12 percent) favor the continued development of nuclear energy, while nuclear power is endorsed by a majority of the European public and an even larger proportion of European Parliament candidates.[11]

The liberal orientation of environmentalists also appears on issues of women's rights. All but one of the environmentalists we surveyed approve of stronger measures for sexual equality. Environmentalists also favor a woman's freedom of choice concerning abortion (88 percent), a percentage slightly higher than the support expressed by CEPs (71 percent). Another survey item shows equally widespread approval (86 percent) of the women's movement. We find clear evidence of the attitudinal sympathy between environmentalists and the women's movement that has characterized other studies of NSM activists (Kitschelt and Hellemans 1990, 113–16; Klandermans 1990; Kriesi 1988).

Within the realm of foreign policy, specifically defense, there is a striking contrast in the opinions of environmentalists and CEPs. Only 22 percent of environmentalists support a stronger European defense effort, whereas 44 percent of the European public and 61 percent of CEPs favor this position.

More than one group representative commented that the cost of a single jet fighter could finance a whole program of underfunded environmental measures. European environmental leaders almost uniformly agree that disarmament of both superpowers is the best way to ensure Europe's security. Furthermore, approval of groups active in the peace movement is slightly higher among environmentalists (81 percent) than among the European public (70 percent). We expect that these differences may have moderated in the aftermath of the collapse of the Soviet Union and the democratization of Eastern Europe, but we also expect that the environmental movement will remain a critic of the defense establishment and continue to prod for even greater cuts in defense budgets and even sharper cutbacks in troop strengths.

Our findings suggest that a distinct New Left identity exists among the leaders of Europe's major environmental groups—which in turn suggests that the leadership of the environmental movement has not succumbed to the moderating forces implied by the Weber-Michels model of organizational development—at least not yet. On the issue of New Politics that are the core of the movement, environmental activists hold strong and distinct views.

Moreover, this New Left identity is especially pronounced among the representatives of ecology groups—indeed, they are the vanguard of the New Politics.[12] For instance, although 41 percent of the respondents from conservation groups are postmaterialists, all of the ecologists choose postmaterial goals; although 21 percent of the former group strongly oppose the development of nuclear energy to meet future needs, 75 percent of the latter group oppose nuclear power. Ecology groups are much more willing to take up the concerns of New Politics, and their leadership are distinctly more supportive of these policies.

Many representatives from conservation groups also subscribe to New Politics, but their opinions are more moderate and they recognize that their membership is even more centrist. Most traditional conservation and nature protection organizations remain hesitant to take an official stand on such issues as nuclear power and disarmament because of possible disagreements among their members. Our interviews found that even when the representatives of conservation groups supported New Politics positions in the simple agree/disagree format of our survey, they added qualifiers that moderated their opinions: disarmament was favored, but it must be phased and balanced; abortion rights were favored but with some conditions; nuclear power was problematic but perhaps it was a necessary evil. Nature protection societies, we were frequently told, are concerned with birds and other animal species— not energy policy and nuclear weapons. Given the political resources that conservation groups control, this political moderation inevitably lessens their potential support for New Politics. Still, there appears to be greater political compatibility between these different elements of the movement at the elite level—and thus the potential for informal political cooperation—than there is between their respective constituencies.

Our findings also differ from past research on green party activists, which argues that environmentalists have left-libertarian views, combining Old Left and New Left political beliefs (Kitschelt 1989; Kitschelt and Hellemans 1990). We find little evidence of the former but clear evidence of the latter. The environmentalists we sampled appear ambivalent about the traditional economic conflicts of European politics, although they strongly support the agenda of New Politics. In terms of the traditional Left/Right economic cleavage of European politics, the movement overall is neither Left nor Right. This is not altogether a surprising finding. It is easy to explain how activists in green political parties develop leftist ideological positions in conflicts between New Politics and Old Politics, because they must address both issues in developing party platforms and election campaign strategies. But this implication is often missed in party-based research. Our findings thus offer a balance to past research—party activists are not representative of all elements of the environmental movement.

Orientations toward the Political System

Some of the sharpest criticisms aimed at environmentalists focus on their supposed opposition to conventional political methods and, by inference, democratic political norms. Defenders of the political status quo frequently characterize groups that promote direct action methods and challenge the dominant social paradigm as environmental extremists and a societal threat. Indeed, there are revolutionary elements within the environmental movement, and the violent actions of some activists—such as the autonomous groups and antinuclear-power protesters in Germany—raise the specter of antisystem opposition. Many visible figures in the environmental movement also criticize the mechanisms of democratic majoritarian decision making as contributing to the environmental problems facing Western societies (Roth and Rucht 1987; Bahro 1984; Porritt 1984; Manes 1990). These antisystem images of environmental action naturally attract public and media interest, and critical reviews of NSMs and green parties often highlight these elements of the movement (Langguth 1986; Fogt 1987).

Our interviews conveyed a much different image. The leaders of most of Europe's major environmental interest groups are not revolutionaries but reformers. Discussions of important issues, government actions, and future

TABLE 5.3

Attitudes of Environmental Elites, the European Public, and Europarliament Candidates toward Political System

	Environmental Elites	Europarliament Candidates	European Public
Satisfaction with democracy			
Very	2%	—	7%
Fairly	64	—	50
Not very	30	—	30
Not at all	_5_	—	_13_
Total	101%		100%
Social change desired			
Need revolutionary change	0	5	5
Need massive and rapid reform	30	29	} 65
Introduce gradual reforms	66	59	
Defend status quo against subversives	_5_	_7_	_30_
Total	101%	100%	100%

Sources: 1979 Study of Candidates to the European Parliament; public opinion data are from Eurobarometer 25 (March 1986), weighted to be representative of the European public. The item concerning satisfaction with democracy was not available in the Eurocandidate study.

political needs nearly always reflect a belief that political reform must occur within the existing system. When environmentalists call for a change in the nature of society, they think of making the public and government more sensitive to environmental issues and perhaps developing a more sustainable style of living. Environmentalists' criticisms of the political process often had the tone of individuals wanting the system to empower the public further and to live up to its democratic ideals. We met many idealists but few radical ideologues.

Several of the questions from our survey illustrate this point. When asked to judge their satisfaction with the way democracy works in their country, 66 percent of environmentalists are fairly satisfied or more with the functioning of the political system; in comparison, only 57 percent of the European public are as satisfied (table 5.3)! Another question asked how much social change was desired. Not a single environmentalist called for revolutionary change, whereas at least a trace of revolutionary sentiment exists among the European public and political elites. The environmentalists we surveyed prefer social reform and differ only in the pace and scope of that change. Positive views about the political system do not mean that environmentalists are uncritical of the political process; in subsequent chapters we shall recount how they are heavily critical of government policy and the environmental records of most political actors. Criticism of the government's environmental record, however, is not translated into criticism of the political process, and that is the important conclusion to draw from these findings.

Part of the reason why the leaders of Europe's environmental interest groups appear more moderate in their political views is that they represent a different constituency than the militants of green parties. Most environmental interest groups attract politically moderate members who advocate environmental reform (Cotgrove 1982; Milbrath 1984). In addition, the findings reflect the mix of environmental interest groups included in our study. Frankly, few political analysts claim that bird societies or cultural preservation groups are revolutionary organizations. Ecologists are the supposed advocates of a New Environmental Paradigm and, as a result, critics of the establishment.

Indeed, we find that environmental orientations are strongly related to reformist sentiments. The leadership of ecology groups are distinctly more critical than conservation groups of social and political systems (fig. 5.3). Two-thirds of the representatives of ecology groups express dissatisfaction with the way democracy functions in their nation, compared to only 12 percent of respondents from conservation groups. Nearly half of the former group claim rapid and massive social reform is needed compared to only one-tenth of the latter. Yet, even those ecologists who advocate massive and rapid social reform often spontaneously add that the focus of these reforms should be on specific environmental needs rather than on a broad restructuring of society. The head of one group cited a package of legislation that was needed, another mentioned the need to increase recycling, some complained about the political influence of business (or agriculture). Overall, the discussions lacked the calls for revolutionary upheaval or fundamental social change that are often highlighted in the political treatises of well-known green activists.

Subsequent chapters will document how the style in which ecology groups frame their political objectives and their strategies of action reflect the reformist values of their leadership. Ecology groups and their leadership

FIGURE 5.3
Attitudes toward Social and Political System, by Group Type

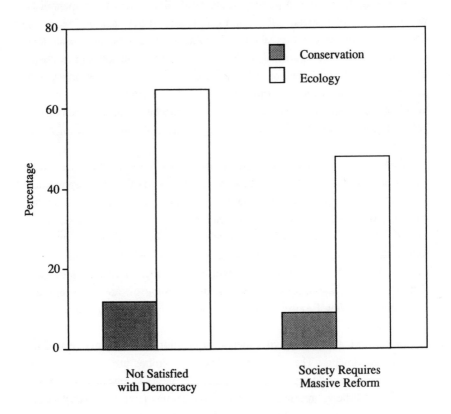

represent a challenge to contemporary political systems, but it is a challenge of political reform and not of revolution.

Middle-Class Radicalism?

The answer to the question of whether environmentalists are extremists or middle-class radicals depends on the definition of *radical*. We do know that the leadership of the environmental movement is socially unrepresentative of the general European public. Environmentalists are disproportionately drawn from the new middle class, which implies a knowledge of organizational skills and familiarity with the manipulation of political symbols. We would stop short of the conclusion others have made, however, which claims the class bias of environmentalism is evidence of middle-class radicalism. It is a middle-class movement, but is it radical? In fact, its social base might lead one to expect just the opposite, because the people we interviewed have obviously benefited from the existing social order. They have risen to the top rungs of the educational ladder and have personally gained from the socioeconomic progress of contemporary society, even if they work outside of the establishment.

Radicalism is better measured by political beliefs than by social characteristics, and here too the evidence is mixed. Environmentalists accept the welfare state consensus of European societies and are, in fact, slightly more moderate than the public in endorsing the classic issues of the Old Left, such as government management of the economy and the reduction of income inequality. On the economic issues that have traditionally structured Left/Right political competition within Europe, the leaders of these major environmental interest groups are far from radical.

At the same time, environmentalists—and especially ecologists—believe that the political agenda should broaden to include issues of New Politics, such as environmental quality, citizen participation, alternative energy sources, sexual equality, and greater international cooperation. Although not revolutionary, these policies still imply basic changes for European societies. The rejection of unlimited economic growth in favor of environmentally conscious growth, for example, can alter the Western economic system in fundamental ways. Similarly, political observers no longer describe the women's movement in revolutionary terms, but the transformation of gender relations in contemporary societies is having profound effects on social relations, employment patterns, and life-styles. According to Ronald Inglehard (1977), these are the issues that are stimulating a change—a "silent revolution"—in the politics and social life of advanced industrial societies.

Most environmentalists also are products of the pluralism and democratization of postwar Europe. Although they generally criticize their government's performance on environmental issues, they nevertheless display high levels of system support and want greater citizen involvement in the political process. Criticism of the political system is more common among the leadership of ecology groups, but even here the intensity of the criticism is muted. Ecologists do not call for revolutionary change but for fundamental reform within the existing political order. These environmentalists exemplify what Samuel Huntington (1981) calls a spirit of "creedal passion," a belief that Western democracies should live up to their democratic rhetoric. These political orientations represent a challenge to the status quo, but it is a challenge evolving from and within the past accomplishments of Western democracies.

A final point has relevance to the analyses in subsequent chapters. Previously we categorized environmental groups in terms of their environmental orientations; in this chapter we described the political orientations of individual environmentalists. We find a large, though imperfect, overlap between organizational and personal characteristics. As we expected, ecology groups attract individuals who are more leftist, more postmaterialist, and less satisfied with the existing political order. These are the adherents of the New Environmental Paradigm and the vanguard of any reformist element of the green movement. The representatives of conservation groups are less liberal in their opinions and more supportive of the political system. The task now before us is to determine whether these two shades of green affect the perspectives and behavior of their respective environmental organizations.

4

POLITICAL REPERTOIRES

Chapter Six

DEFINING THE

AGENDA

In the months preceding our survey of European environmental groups, a reader of the *Times* of London would have learned an extraordinary amount about the environmental issues affecting Britain, Europe, and the rest of the world. Numerous stories addressed the protection of birds, from the dedication of new preserves to reports of hunters who had been fined for killing protected species. Citizen protests ranged from one individual erecting a cross to mark a farmer's destruction of a three-hundred-year-old hedgerow to mass demonstrations organized by FoE against government policy on nuclear power. Some stories were humorous, such as one about the man who lived with a whooping crane to encourage it to lay eggs; more often the reports were distressing, such as stories of increased cancer among the children of workers at the Sellafield nuclear power plant. Indeed, scores of articles dealt with the problems of Sellafield, the three-year inquiry into plans for the Sizewell plant, and other issues of nuclear power and waste disposal. Acid rain was another prominent topic, eliciting a virtually continuous flow of stories concerning scientific evidence on damage to the environment and the international tensions created by Britain's obstinacy on the issue. Other articles discussed problems such as rising levels of toxic metals in water supplies, chemical spills and pesticide contamination, and the resurfacing of Westminster Abbey because of air pollution. The Conservative, Labour, and Alliance parties made regular policy statements on environmental issues, which often prompted speculation on the need and prospects for a green party in Britain.

Even this one newspaper detailing the problems in one nation illustrates the diversity and complexity of green politics. Providing a more complete definition of the green agenda is an important first step in studying environmental action. Europe now faces a host of environmental problems that require attention. The views of group representatives can define which are the most important problems for governments to address—at least for those individuals who are most interested in these issues.

We also expect to find that a group's environmental orientations influence its perception of issues and possible activities. At a basic level, a group's perceptions of the important national environmental problems will influence judgments about which arenas of action they consider appropriate and perhaps which tactics are suitable. Similarly, a group's own issue priorities determine the nature and scope of the problems it faces in attempting to affect policy change and thus its choice of tactics and likelihood for success (Gamson 1975, chap. 4). A group concerned with reducing acid rain obviously confronts different options and challenges than a group intent on protecting the English hedgehog.

Furthermore, if our model of Ideologically Structured Action is correct, judgments about the content of the environmental agenda will be colored by the political orientation of a group. David Snow and Robert Benford (1988; Snow et al. 1986) describe this process as "diagnostic framing." The ideology of an SMO affects its identification of the most salient problems: which environmental problem should green groups address first, or what aspect of sexual inequality should the women's movement first address? Even when a consensus on problem identification exists—such as with the peace movement during the controversy over NATO missiles in the early 1980s—ideology can still strongly influence the attribution of blame or causality.

Because of their more fundamental view of environmentalism, we expect ecology groups to frame a set of national problems (as well as group goals) that differ from the agenda of conservation groups—even if we ask them to separate their group's own priorities from the national agenda. Not only should ecology organizations differ from conservationists in their perceptions of the content of the environmental agenda, but diagnostic framing should also lead them to hold different perceptions of the methods of addressing these issues and the impediments to policy success. The variation among our environmental groups can test this aspect of ideologically structured behavior. If our ideas are supported, perceptions of the green issue agenda may, in turn, precondition the political views and behavior of these organizations.

This chapter provides a brief introduction to the policy orientations of European environmental groups by asking them to assess the important environmental problems in their nation, as well as the problems of concern to their own organization. We attempt not only to define the agenda for environmental action but to use these issues to illustrate how groups perceive the political context in which environmental reform must be achieved.

Issue Priorities

At the time of our survey, environmental issues had become prevalent in most European nations. Some issues, such as acid rain and the consequent destruction of forests and lakes, were receiving international attention. Through a series of governmental conferences and negotiations, the European Community member states were in the process of addressing the acid rain problem (Vogel 1988, 1993). The death of the forests in Germany and Scandinavia was an undeniable example of the impending crisis that environmentalists had warned about. The 1980s were also a decade of debate and conflict over nuclear power, an issue that touches most of the nations in our study (Hatch 1986; Rüdig 1990).[1] Issues of wildlife protection were attracting broad interest across Europe; treaties restricting the trade of endangered species brought new attention to the issue, as did the dramatic Greenpeace campaigns to protect whales and baby harp seals. In addition to these well-publicized international problems, many other nation-specific issues called for urgent action.

To determine how environmental groups define the major contemporary issues, we asked group representatives which problems they would place at the top of their nation's environmental agenda.[2] The more exotic environmental issues often capture the greatest media attention, but the more mundane issues of countryside protection and agriculture actually generate the most concern among environmentalists (table 6.1). A full 71 percent of all group representatives mention countryside issues as a national priority. The heavy urbanization and industrialization of Europe have created a need to protect the remaining natural environment. One British respondent described the sentimental attachment of the British to the countryside; to many the idyllic landscape composed of small farms and hedgerows represents Britain's heritage, and large-scale farming or economic development projects upset these images. Nearly every British environmental group mentions this concern for the countryside.

The British, however, are not the only Europeans who are attached to the landscape. Several of the German environmental groups describe similar feelings toward the national forests; many Dutch and Danish groups express a general concern with preservation of the countryside. In some cases, large

TABLE 6.1

Perceptions of Important National Environmental Problems

	Conservation Groups	Ecology Groups	Total
Preservation issues			
Countryside and agriculture	91%	47%	71%
Wildlife protection	26	18	24
Urban renewal/planning	12	12	16
Advanced industrial issues			
Air pollution/acid rain	53	82	62
Water pollution/supply	44	65	52
Chemical waste	12	24	16
Solid waste disposal	15	18	14
Nuclear power/waste disposal	6	24	12
Energy alternatives/problems	6	25	10
Other problems	41	35	38
Total	306%	349%	315%
(*N*)	(34)	(17)	(58)

Note: Figures are the percentage of group representatives naming the given area as a problem. Totals exceed 100 percent because up to five responses were possible. Groups with a mixed environmental orientation are not listed separately in the table.

development projects are the threat. For instance, major airport expansions in Britain (Stansted) and Germany (Frankfurt's Startbahn West) generated headlines in the 1980s. The problems of agricultural overproduction, often stimulated by the policies of the European Community, are another frequent source of environmental concern. On the whole, the protection of the remaining countryside and other natural areas looms as a top priority for most environmental groups.

Acid rain is the most frequently cited environmental problem, mentioned by one-third of the sample. Moreover, acid rain is a truly international issue; expressions of concern range from Ireland to the Mediterranean. German groups were especially vocal about the issue because they view themselves as a net recipient of other nations' pollution, but concern with acid rain is also common among environmental groups in nations creating the pollution, such as Britain. Not only does the issue of acid rain transcend national borders, but group representatives believe that only international action can resolve this problem. Another air quality problem is of special concern in Greece; every Greek group mentions the choking smog (*nephos*) of Athens as a major environmental worry.

One group representative describes water pollution as the "sleeping environmental crisis" in Europe—a view shared by many. Half of the environmentalists mention some water problem as a national priority, but the nature of the issue varies by country. In agricultural nations such as Denmark, there is widespread concern with the problems of agricultural runoff; several French groups cite problems in guaranteeing adequate water supplies; industrial pollution is the prime concern in other nations.

Beyond the three priority areas—protection of the countryside, air pollution, and water quality—numerous other issues receive moderate attention.[3] Only one-quarter of the sample cite wildlife protection issues, even though the study contains many conservation groups. Other problems include chemical and industrial wastes, solid waste disposal, urban renewal, and energy issues.

Perhaps most surprisingly, the issue of nuclear power is nearly missing from this list. Despite the high visibility of the nuclear power controversy in Europe, barely one-tenth of the groups cite this as an important national issue. Even where national governments were actively discussing a nuclear power program (e.g., France, Belgium, Italy, and Germany), the same small percentage mentions nuclear power as a pressing issue. More often than not, environmental groups are hesitant to address the nuclear power issue in formal group terms, because it tends to divide their membership. Rather than a vanguard issue for the environmental movement, nuclear power registers as only a small part of a much larger and more varied political agenda.

A group's environmental ideology at least partially determines its perceptions of which environmental issues are most important for the nation (table 6.1). In keeping with their preservationist orientation, conservation groups emphasize issues such as countryside preservation and protection of wildlife. These are generally consensual political issues, for which conservation groups attempt to mobilize existing popular support. Indeed, issues of just this sort led Lowe and Goyder to label most conservation organizations as "emphasis groups" that stress popular issues politicians might otherwise overlook in the ongoing competition between political interests (1983, 35).

Ecology groups, in contrast, espouse their own agenda of issues. Ecologists are more often concerned with what we term advanced industrial issues—excessive growth, new technologies (chemical and nuclear wastes), and industrial pollution—and are more likely than conservation groups to cite these as important national problems. For instance, 82 percent of ecology groups cite air pollution as an important national problem, as opposed to 53 percent of conservation groups. Moreover, these issues inevitably lead ecology groups to challenge the industrial and life-style patterns responsible for water pollution problems, acid rain, toxic wastes, and nuclear power.

The different patterns of response by ecology and conservation groups are, perhaps, more similar than they first appear. A conservation group might express a concern with the destruction of forests, whereas an ecology group might describe the same problem in terms of air pollution and, more specifically, acid rain. The different frames of reference in this example are significant: the goal of one group is to protect a specific natural resource; the other group focuses on the industrial and automobile pollution that produce acid rain. When viewed in these different ways, the same problem can lead to alternative policy goals and prospects for action—this is the importance of diagnostic framing.[4]

The contrast between preservation issues and advanced industrial issues is also represented in the policy goals of the groups themselves. We asked group officials to list the major issues their group had addressed in the recent past (table 6.2).[5] Group activities on issues of nature conservation and preservation are even more common than the ratings the groups had given these issues as important national problems (cf. table 6.1). In part, this is because conservation groups greatly outnumber ecology groups in our sample; but even ecology groups are more likely to take action on preservation issues than to identify them as national problems. We believe that both conservation and ecology groups advocate wildlife protection and nature conservation to attract popular and financial support. Environmental organizations find that public interest in wildlife and nature presents an opportunity for discussing broader environmental matters, such as the effects of industrial pollution and agricultural practices. The head of one group admitted that although most of his members are interested only in wildlife, he used this interest to illustrate how bird habitats are threatened by urban development and how the health of wildlife can be endangered by industrial pollution. A by-product of this preservationist emphasis is that advanced industrial issues are underrepresented in the recent activities of environmental groups. The number of groups addressing the potentially contentious advanced industrial issues is generally smaller than the number of groups that cite these as important national problems. Ecology groups are the strongest advocates of

TABLE 6.2

Most Important Environmental Problems Addressed in Recent Past

	Conservation Groups	Ecology Groups	Total
Preservation issues			
Countryside and agriculture	90%	48%	74%
Wildlife protection	64	38	49
Urban renewal/planning	18	5	16
Advanced industrial issues			
Air pollution/acid rain	26	48	35
Water pollution/supply	10	19	19
Chemical waste	13	38	21
Solid waste disposal	8	10	9
Nuclear power/waste disposal	8	38	19
Energy alternatives/problems	10	25	16
Other problems	13	27	22
Total	260%	296%	280%
(*N*)	(39)	(21)	(68)

Note: Figures are the percentage of group representatives naming a problem in that area. Totals exceed 100 percent because up to five responses were possible. Groups with a mixed environmental orientation are not listed separately in the table.

these issues, but even they devote much of their energy toward traditional preservationist goals.

What Is to Be Done?

Defining the problem is but the first step in the process of environmental action. Having identified important national issues, an individual or organization must also determine the underlying source of the problem, the factors standing in the way of policy reform, and the actions required to solve the problem. In part, the nature of an issue affects how the political context for environmental reform is perceived. Some issues might require merely strict adherence to prevailing social or legislative norms—as in the case of endangered species—and the primary impediment to action is the inertia that confronts any new policy proposal. Other issues, such as nuclear power, reflect the new socioeconomic conditions of advanced industrial societies. Their resolution could require a more fundamental restructuring of the socioeconomic system, such as reversing the government's research and development priorities in the energy field and shifting to decentralized methods of power generation.

Evaluations of the context in which to deal with an issue are not politically neutral. Diagnostic framing also occurs when the political beliefs and values of an environmental group influence perceptions of the underlying source of a problem and the steps required to resolve it. What one group sees as a problem of insufficient legislation, another may consider a consequence of the distribution of power in society. Views on the roots of environmental problems may therefore provide a clearer insight into the political ideology of the environmental movement than the simple question of which problems are most important.

In our interviews, we asked respondents to discuss the political context for addressing the issue cited as the most important environmental problem facing their nation (table 6.1). We then asked group representatives what they thought must be done to solve this national problem (table 6.3). The picture that emerges is of a movement engaged in policy reform rather than in a fundamental restructuring of society. A majority of groups (59 percent) feel that their nation's most pressing environmental problem can be resolved through the strengthening of existing legislation or the passage of new legislation. The tenor of these responses suggests that radical social restructuring is not at issue. Most calls for legislative initiatives focus on such matters as enacting stronger emissions standards, tightening enforcement of existing legislation, and enacting legislation that has already been accepted in other European states. Moreover, these are not casual calls for legislative action. Our interviews clearly revealed that many of these organizations were deeply involved in the legislative process, in the form of conducting research, drafting legislative proposals, and directly working with supportive parliamentarians. Ecology groups place a slightly greater stress on legislative reform than do conservation groups, presumably because their policy agenda focuses on new environmental problems requiring new legislation. By comparison, as emphasis groups, conservationists are more likely to feel that their needs can be met by pressuring the government into action; they know that for many issues the legislation and popular mandate for action already exist.

TABLE 6.3

Solutions to National Environmental Problems

	Conservation Groups	Ecology Groups	Total
Strengthen/enact legislation	53%	76%	59%
Influence public opinion	43	59	48
Pressure politicians/government	33	18	30
Fundamental social change	13	24	24
Develop/use technology	13	35	22
Financial solutions	23	24	20
Other solutions	27	24	30
Total	205%	260%	233%
(*N*)	(30)	(17)	(54)

Note: Figures are the percentage of groups naming a solution in the given area for the problem listed as most important in table 6.1. Totals exceed 100 percent because up to five responses were possible. Groups with a mixed environmental orientation are not listed separately in the table.

Environmental groups place a high priority on public opinion as a stimulus for policy reform. They emphasize the need to develop greater public awareness of issues and to mobilize existing popular support for reform rather than to change public values substantially. We also should note that the data contradict a popular myth that environmentalism is a backlash against science and technology. About one-fifth of the groups spontaneously mention the development or utilization of technology as a remedy for environmental problems; surprisingly, such responses are more frequent among ecologists.

Again, what is missing from the data presented is evidence of a strong radical or revolutionary orientation among environmental interest groups. Strikingly few groups (24 percent) mention a need for fundamental social change as a prerequisite for environmental reform, including ecology organizations. Even among most of the groups that advocate basic social change, the details of their answers weaken the force of their responses. Often, the reference to social change involves a vague call for a shift in life-style or in the relationship between energy and economics. In other instances, the apparent desire for "basic" political change is essentially reformist; two Belgian groups, for example, advocate a reorganization of the country's federal system to address the nation's environmental problems. The moderate tone of these calls for "fundamental" social change is also illustrated by the views of one Greenpeace group on the problem of chemical wastes from industry. The GP representative went into great detail on the research and testing that had been done by Greenpeace experts to assess the problem, the efforts they had taken to publicize the issue, and their assessment of the specific flaws in existing legislation. Greenpeace saw the primary route for change as enacting new legislation, and reference to the wider social aspects of the problem were mentioned almost as an aside. In our discussions, the environmental lobby falls far short of the radical rhetoric that is so prominent among many green politicians and academic theorists.

As we noted above, prescriptions for action are partially dependent on the specific issue under discussion. Groups that focus on wildlife protection issues are less likely to see fundamental social change as necessary to address this problem. Overall, however, the choice of issue has only a weak effect on the prescriptions for action proposed by a group. The differences between conservation and ecology groups in perceptions of what must be done are generally larger than the differences produced by issue type. This reinforces the point that environmental orientation, rather than the issue itself, is more important in framing the policy perceptions of an environmental interest group.

Despite the modest reformist goals of most environmental groups, this does not mean that they are sanguine about their prospects for success. In fact, when asked if the most important national problem will be solved in the next five years, only 6 percent of the groups are optimistic (table 6.4). The greatest perceived impediments to success come from politicians, political parties, and government agencies. Our interviews are replete with stories about negative experiences in dealing with the government; in most cases the stress is on the lack of interest or support by politicians. One environmentalist claimed that the only difference between the Ministry of Industry and the Ministry of the Environment was that the former was the Ministry of Causes and the latter was the Ministry of Effects; otherwise their political viewpoints were often indistinguishable. In other cases, groups had experienced outright hostility by government agencies. French and Greek groups feel betrayed by the socialist governments they supported in the prior election, and British groups confronted a government that was unresponsive to their interests.

Group representatives also see that their attempts at environmental reform face formidable opposition from other socioeconomic interests. Issues of industrial pollution and limits on economic development generate resistance

TABLE 6.4

Will Important National Environmental Problems Be Solved?

	Conservation Groups	Ecology Groups	Total
Will be solved	3%	13%	6%
Political impediments			
Lack of interest/opposition by government and parties	48	31	46
Need more legislation	6	13	9
Systemic impediments			
Opposition from social groups	29	56	35
Requires change in political or socioeconomic system	10	25	19
Other impediments			
Just takes time	45	13	31
Need more public support	10	25	15
Financial reasons	13	6	11
Other reasons	6	6	7
Total	170%	188%	179%
(*N*)	(31)	(16)	(54)

Note: Group representatives were asked whether or not they believed the most important problem in table 6.1 would be solved in the next five years. If they answered no, they were asked to name the major impediments to its solution. Totals exceed 100 percent because up to three responses were possible. Groups with a mixed environmental orientation are not listed separately in the table.

from a loose alliance that unites business and labor interests. Business interests normally take the lead in opposing these policies of environmental reform, but labor unions are also cited as an impediment to success. The opposition of agricultural interests to countryside issues is even more apparent. More than one of our respondents stated that agricultural ministries and the farming community were producing the greatest harm to Europe's environment. Agricultural interests have been vocal opponents to proposals that would restructure farming practices with different land-use incentives, restrict the use of fertilizers and pesticides, and reform EC subsidies that encourage overproduction. In short, environmentalists are clearly aware that although the public generally supports their proposals, major economic interest groups are often adversaries.

Our discussions about the prospects for environmental reform also illustrate differences in the Weltanschauung of conservationist and ecology organizations. Many conservation groups (45 percent) believe that given the gradual changes that are occurring, resolution of their nation's environmental problems is a question of more time; far fewer ecology organizations (13 percent) perceive such a simple dynamic at work. Ecology groups disproportionately see systemic factors—such as opposition from social groups or the need for change in the socioeconomic system—as restricting environmental reform. Ecologists also give a more prominent role to public opinion, placing more emphasis on the need to develop greater public attention to achieve success. Ecology groups realize that their interests represent a greater challenge to contemporary society, albeit a modest challenge.

A Movement of Reform

As a first introduction to the policy goals of environmental interest groups, this chapter illustrates the diversity of interests that are included under the umbrella of green politics. Environmentalists are concerned with a complex and multifaceted set of political issues. And in discussing these issues, they display an awareness of the intricate interactions between problems of industrial pollution, nature preservation, wildlife protection, and preservation of the quality of life. Bird societies recognize that chemical wastes and agricultural overproduction will threaten the sanctuary of their nature preserves. Ecology groups in Europe believe preservation of tropical rain forests to be as important as protection of the European forests from acid rain.

Our data also present a picture that differs from the image of environmentalists as radicals intent on restructuring society. In terms of their issue interests, most group representatives see the most pressing problems as involving modest issues of environmental reform, a stance that echoes their personal political values. Certainly, contentious issues such as nuclear power and toxic waste constitute important concerns for these groups, but issues of a more moderate character make up much of the environmental agenda. The destruction of the natural environment, the effects of agricultural overproduction, and the protection of wildlife actually receive the most attention from the environmental lobby. Furthermore, in their prescriptions for environmental reform, most groups emphasize the need for broadening legislation, pressuring government officials, and mobilizing public opinion; calls for radical societal change are far less frequent. In short, environmental interest groups do not see themselves as pitched in a fundamental confrontation with the contemporary sociopolitical order; rather, they are addressing a policy area that already has broad public backing.

Finally, our findings highlight the importance of a group's environmental orientations in structuring its perceptions of the political world and in providing a framework for organizing and interpreting political events. Conservation groups emphasize issues of nature and wildlife protection; ecology groups are more likely to advocate the new issues of advanced industrial societies, such as nuclear power, acid rain, and toxic wastes.

Even more important, environmental orientations influence the way a group diagnoses an issue and identifies the remedies. As others have recently argued (Ferree and Miller 1985; Snow and Benford 1988; Klandermans 1991, 8–10), a group's identity or political orientation provides it with a construction of the political world. Often these perceptions of reality, combined with reality itself, guide political actions. Thus, even when addressing similar political issues—for instance, wildlife preservation issues—conservation and ecology groups advocate different courses of action. Conservation groups often view environmental reform as an evolutionary process of policy change, whereas ecology groups realize that many of their policy goals may require a broader restructuring of social or economic relations. Conservation groups do not see themselves as outside the established political order; ecology groups are more likely to see business, labor, and other major social actors as opposing their interests.

Environmental orientations thus frame both the definition of the problem and perceptions of possible solutions. In these varied definitions lie the seeds for diversity in goals and actions.

Chapter Seven

ALLIANCE PATTERNS AND

ENVIRONMENTAL NETWORKS

If environmental groups hope to implement policy reforms, they cannot do it alone. Politics today is a process of coalition building—allies must be courted and possible opponents counterbalanced. Identifying these potential friends and allies is a crucial step in building political coalitions and planning strategies.

The significance of such alliance patterns is a matter of contention among social movement scholars. The organizational emphasis of resource mobilization theory focuses on how social movement organizations participate in coalition formation (Curtis and Zurcher 1973; McCarthy and Zald 1977; Tilly 1978; Klandermans 1989). At the individual level, participants in a social movement retain personal ties to labor unions, churches, political parties, or other groups; these multiple affiliations establish a link between a social movement and these other groups. Personal affiliations are important in mobilizing new participants, as well as in affecting the behavior of individual SMO members (Klandermans et al. 1988). For instance, adherents of the environmental movement overlap considerably with supporters of other New Social Movements (NSMs) and are drawn disproportionately from communities with high mobilization potential for alternative political groups (Kaase 1990; Kriesi 1988; Kitschelt 1989).

Social movement scholars also stress the importance of networks at the organizational level: these are the ties that link SMOs to other social groups and political organizations (Klandermans 1990; Diani 1990b). A social movement is itself a network of participants, and this network is in turn linked to other organizational networks through alliance patterns. Charles Tilly's

(1978) distinction between members of the polity and challenging groups (such as social movements) is relevant here. Although not a formal part of the social movement itself, members of the polity can facilitate the development and maintenance of the movement. The contributions of funds, personnel, and expertise from other organizations—such as labor unions, churches, charitable foundations, and political institutions—can be crucial in assisting a movement to overcome the hurdles of creating a new organizational entity. Members of the polity may serve as a conduit for political influence, giving a contentious movement at least indirect access to the policy process and linking SMOs to potential supporters and advocates within the political establishment. The importance of alliance patterns is illustrated by research indicating that the number of organizational allies supporting an SMO is a significant predictor of success (Gamson 1975; Steedly and Foley 1979; Turk and Zucker 1984). RM theory thus sees social movements as embedded in a complex network of organizations—a "network of networks" in Friedhelm Neidhart's (1985) terms—with overlapping interests, personnel, resources, and bases of popular support.

In contrast to the resource mobilization perspective, many of the initial writings on NSMs, such as environmental groups or the women's movement, described these organizations in terms of their estrangement from members of the polity and other established social and political groups (Brand et al. 1986; Offe 1984). Claus Offe attributed the isolation of NSMs to their dichotomous view of society—the conflict between us and them—a view that discourages political exchange or gradualist political tactics (1985, 830).[1] Offe further maintained that NSMs distance themselves from established sociopolitical groups because the popular base of these movements is drawn from the new middle class and other decommodified groups that consciously remain on the periphery of society (851–52). Other scholars link the supposedly autonomous tendencies of NSMs to their lack of resources and hence their inability to engage in the bargaining exchanges of normal politics (Melucci 1980).

The burgeoning literature on NSMs yields mixed support for these contrasting images of their organizational networks. In the United States in particular, environmental groups seem to have quickly integrated themselves into established political networks, using various paths. Private foundations and public grants were important in facilitating the creation of many environmental organizations and other citizen groups (Walker 1991, chap. 5; Downing and Brady 1974). Local citizen groups formed in the wake of the Three Mile Island incident had numerous (and diverse) ties to other political groups (Walsh 1989).[2] In contrast, some environmental groups—such as

Friends of the Earth, Greenpeace, and Earth First!—apparently remain outside the circle of established politics (Manes 1990; Shaiko 1993). There is also an undeniable antiestablishment tone to much of the rhetoric of the green movement in Europe (Rucht 1989; Brand 1982, 1985; Diani and Lodi 1988; Diani 1990a).

As with other aspects of the environmental movement, we expect that perceptions of these alliance networks represent an interaction of objective conditions and political orientation. Government ministries, for example, may be more or less attuned to environmental reform, but environmentalists' evaluations of the ministries will be conditioned by their general orientations toward the political system as well as by the specific performance of the ministries. The willingness of government agencies and other political actors to build alliances with the environmental movement also should depend on the political identity of each group. Furthermore, the differing priorities of environmental groups may lead to different assessments of the objective condition; FoE will judge the environmental ministry on different terms than will a historical preservation society. To illustrate: although the Royal Society for the Protection of Birds (RSPB) and Greenpeace are both concerned with wildlife protection issues, there are predictable differences between the potential resources and political opportunities available to these two groups. A group with a nonchallenging ideology, such as the RSPB, is more likely to receive support and build alliances with government agencies and societal interests that are part of the prevailing social order. A group such as Greenpeace may have a disdain for established groups that will lead it to find allies among other movement groups or dissident factions of society.

In this chapter we evaluate how environmental interest groups relate to other significant political actors in the environmental area. First we address the extent to which these groups are able to identify potential allies among the established social and political interests in their respective societies. We begin by tapping environmentalists' perceptions of the major social and political groups. Which groups are seen as sympathetic, and which as antagonistic? If alliance patterns do exist, what is the composition of this network for the movement? Are environmental groups integrated into existing political networks or do they develop alternative allies? Having defined these general patterns, we then determine the extent to which environmental orientations structure these perceptions. Further analyses examine the variation in perceptions of specific actors, such as labor unions or government agencies, across nations. Finally, because the densest part of this organizational network involves cooperation (or competition) with other members of the en-

vironmental movement, we examine the potential for alliances among the
separate environmental groups.

Perceptions of Alliance and Conflict Networks

It is difficult to provide a single assessment of the real or potential political
bonds that might link environmental groups to the other major actors in
Western European systems. If we examine the ongoing activities of environmental interest groups, we are likely to find that many of the action-oriented
alliances between environmentalists and other political actors are fleeting
affairs. Jeffrey Berry (1984, 202–05) describes coalition formation as a
ubiquitous part of the interest group politics, where kindred spirits are con-
stantly working to form new coalitions as old ones dissolve. A specific issue,
such as waste dumping at sea, may assemble a coalition of environmentalists,
dockworkers, and fishing interests that exists only for this cause and breaks
apart once it is resolved. Similarly, a campaign to prevent the destruction of
tropical rain forests might create a political coalition that is united on this
topic but cannot agree on actions aimed at preventing the destruction of
European forests by acid rain.

In any case, in this chapter we are interested not in alliances that might
develop on a single issue but in the political predispositions that underlie
patterns of political action and define the relationship between the environmental movement and the polity. Beyond the short-term instrumental ties of
a specific political campaign, there are recurring patterns in how other political actors generally respond to calls for environmental reform. These general
patterns of sympathy or response define the organizational field in which the
environmental movement functions (Curtis and Zurcher 1973). Bert Klandermans (1990) points out that this field contains both supporting and opposing sectors. The alliance system consists of organizations that regularly
support the general goals of environmental interest groups; the conflict system
consists of actors who routinely oppose the movement, often including agencies of the state.

Defining the organizational field of environmental groups can clarify
how these groups relate to established social interests and governmental
institutions; knowing the members of their alliance and their conflict systems
also elucidates their potential resources and the forces that block their course.
For instance, if environmental groups choose to distance themselves from
the government and other members of the polity, their claim to represent a
challenging ideology must be given more weight. The question of how
environmental groups relate to traditional Left/Right political alignments can

suggest the movement's potential for realigning the sociopolitical space of advanced industrial democracies.

To look beyond the short-term patterns of specific issue alliances, we assessed the organizational field of environmental groups by asking group representatives to evaluate the environmental performance of other important political actors. That is, rather than construct this field based on formal institutional ties (which are difficult to assess reliably because of the fluid nature of the environmental movement), we measure this field in perceptual or psychological terms. As we have argued, such perceptions create an image of reality that then structures the actions of these groups.

We asked the respondents in our survey to define their allies and opponents according to how well various organizations are addressing their nation's environmental problems.[3] We consciously focused on environmentalists' perceptions of members of the polity: government agencies, socioeconomic interest groups, and the major political parties. Access to the organizational and political resources of these actors would contribute to the development of the environmental movement. Furthermore, the pattern of these relationships is a key factor in theorizing about the nature of the environmentalism.

In analyzing the responses we averaged the environmental ratings of each actor across the combined European sample (fig. 7.1; responses are coded so that 4.0 represents an "excellent" evaluation by all environmental groups and 1.0 represents a consistently "poor" rating). The averages for conservation groups (marked by a darkened circle) and ecology groups (marked by a clear circle) are presently separately; in every instance the European average lies between these two poles. These data enable us to see if there are systematic differences in the rating of political groups that transcend national borders, and they provide a standard against which evaluations of specific actors in each nation can be judged.

Social movements organize to address an issue that is not recognized or accepted by the dominant forces in society, hence the need for the movement. It is therefore not surprising to find that environmentalists rank only two institutions—green/New Left parties and newspapers—above the good/bad midpoint of the scale. European environmental leaders give predominantly negative—and even sharply critical—environmental performance ratings to the wide range of other social and political actors included on the list.

The ministry of the environment (MOE) of the respective nations received relatively favorable ratings; group representatives rank it near the top of the list of institutions, but still below the midpoint of the scale. By comparison, nearly all European environmentalists are sharply critical of the actions of

FIGURE 7.1
Environmental Evaluations of Social and Political Actors

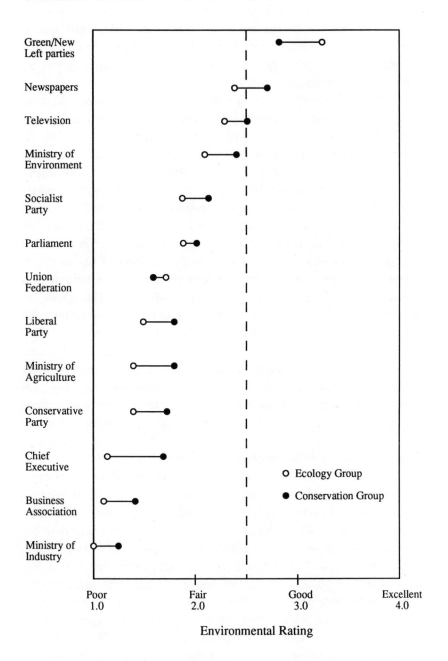

Environmental Rating

their ministry of industry (MOI) and ministry of agriculture (or their equivalents). Environmentalists rank the MOI at the bottom of the scale, often with accompanying comments that even a "poor" classification is too positive. The relative rankings of the MOE and the MOI reflect the policy-making problems of the environmental movement. In the battle between a weak MOE and strong MOI, most of the environmental leaders (71 percent) feel the ministry of the environment will lose.

A similar imbalance exists in evaluations of parliament and the chief executive. The respective national parliaments generally receive an environmental rating above the average of other institutions (mean = 1.96). In contrast, the chief executive consistently garners negative evaluations (mean = 1.43), regardless of whether the Left or Right controls the government. In an age where the power of the executive is predominant, this pattern cannot be considered a positive sign for potential environmental reform.

When one scans the ranking of all the actors in figure 7.1, one sees the shadow of a familiar pattern. Although environmentalists often boast that they represent a new political movement that is neither Left nor Right, perceptions of the allies and opponents of environmental reform at least partially reflect a traditional Left/Right political alignment. Throughout Europe, environmentalists are consistently critical of business, agriculture, and industry in their nation. Nearly 90 percent of our respondents assign their ministry of industry a "poor" rating (mean = 1.11); and more than 75 percent give similar scores to business associations (mean = 1.27). Environmentalists apparently have little difficulty identifying these foes. By comparison, environmentalists rate labor unions more positively than business or agricultural interests, though unions are still not seen as receptive to environmental reform. Almost half of all environmentalists give labor union federations a "poor" rating for their environmental performance (mean = 1.70).

This modest Left/Right pattern in performance evaluations also emerges for evaluations of party groups across Europe (see also chap. 9). Beyond the positive support for green and New Left parties, most environmentalists give negative ratings to all of the established parties, but modest differences emerge between the major party groups.[4] Group representatives rate socialist parties (mean = 2.02) slightly higher than liberal parties (mean = 1.66), and conservative parties a notch lower (mean = 1.55). Averaged across ten separate party systems, the differences among these three established party groups follow a Left/Right pattern, but the largest gap still separates *all* the established parties from the green/New Left parties. Certainly, labor leaders or business executives would display far sharper differences in their perceptions of the established parties' economic policy positions.

These patterns also illustrate how environmental orientations affect perceptions of the organizational field. The ranking of actors is relatively similar between conservation groups and ecology groups. Conservation groups, however, are almost always more positive in their evaluations of social and political actors, in part because established institutions are more receptive to conservation issues. The ideological coloring is especially strong for institutions identified with the dominant industrial order: the prime minister (tau-c = .30), business associations (.23), the ministry of industry (.21), and the conservative party (.23). There are two notable exceptions where this general pattern is reversed and ecology groups are more positive in their evaluations: green/New Left parties (−.23) and labor unions (−.02). This first correlation suggests that ecology groups are more likely to view green parties as the parliamentary arm of the movement. The second correlation hints at the potential for Old Left/New Left alliances, though ecologists and conservationists are still rather critical of the unions' environmental stance.

These evaluations based on the total sample of environmental groups are useful for identifying regularities in political evaluations that transcend national boundaries. Still, we must be cautious about these data because they aggregate evaluations of different national actors. Environmentalists are responding to different objects in each nation. For instance, ratings of the chief executive intermix opinions about Papandreou, Mitterrand, Thatcher, Kohl, and the other European heads of state; such forced commonality may be inappropriate. We therefore want to go beyond these generalized results to look at national patterns.

Governmental Institutions

The relations between a social movement and government agencies strongly affect the political potential of the movement. Finding allies within the government can facilitate the mobilization of resources and access to the decision-making process. Strong opponents within the government—such as antagonistic agencies or political parties—can drain the resources of the movement and restrict its opportunities for influence.

We examined three key government institutions in each nation: the ministry of the environment, the parliament, and the chief executive (fig. 7.2). One should remember that these data are based on a small number of groups responding in each nation (Ireland and Luxembourg are excluded because only two organizations were contacted in both nations). At the same time, the groups we surveyed do represent the major environmental actors in each nation and nearly the universe of major environmental interest groups

FIGURE 7.2

Environmental Evaluations of Political Institutions, by Country

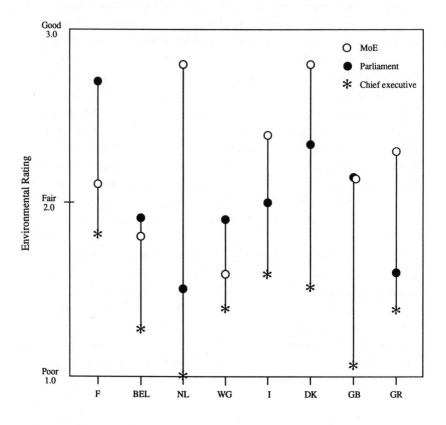

in Europe; their responses are therefore important evidence of the perceptual field of the movement.

Among government institutions, environmental groups generally give their highest ratings to their respective MOE.[5] The MOE is the supposed protector of the environment; if green groups are to find an ally within the government, it should be here. And yet, the absolute rating of environmental ministries was quite modest; only three—Denmark, Holland, and France— receive ratings above the midpoint. Throughout the interviews, we heard repeatedly that relations between environmentalists and the MOE often have an adversarial tone, falling far short of the normal neocorporatist arrangement that might be found between the transport industry and the ministry of transportation, or farmers and the ministry of agriculture. One senior official of Britain's Department of the Environment observed that his was the only government department where the civil servants were expected to work with a clientele that is being continually criticized by the leadership of that department. Indeed, in most nations environmentalists cite the MOE for its inaction or poor relations with green organizations. At the time of our survey, these relations were extremely strained in Germany, where the Ministry of the Interior handled environmental concerns. The conflict between environmental issues and the ministry's other concerns, as well as the conservative orientation of Interior Minister Friedrich Zimmermann, may explain why the German government established a separate Ministry for the Environment and Nuclear Safety in the year following our survey.

A complementary view of governmental performance on environmental issues is available from a Eurobarometer survey.[6] We used this to compare the public's rating of the government's environmental efforts (arrayed along the horizontal axis) with environmentalists' evaluations of the ministry of the environment (along the vertical axis, fig. 7.3). Because our group survey and the Eurobarometer use different questions, we compare only the general positioning of nations, not the absolute ratings across studies. There was an apparent consensus on the positive performance of the MOE in three nations—Denmark, the Netherlands, and France—which are ranked first, second, and third in both studies. Conversely, environmental officials receive low marks from both samples in Belgium and Great Britain. Popular support for the environmental movement was growing in both nations in the late 1980s as a result of government neglect of green issues, and green parties made their strongest two showings in the 1989 European parliamentary elections in Belgium (13.9 percent) and Britain (14.5 percent). Finally, there are two instances in which the opinions of environmentalists and the public diverge. The German public is more positive in its evaluation of the government's general environmental record than are German environmentalists. We suspect this reflects the transformation of government policy that was occurring under Kohl, including the creation of a new environmental ministry in June 1986. The Italian public is most critical of its government's environmental performance, but this results more from a lack of concern for the problem than from negative actions (Commission of the European Communities 1986). Across all eight nations, however, there is a marked similarity in the governmental evaluations of European publics and environmentalists.

Environmentalists normally rank parliament second among these three institutions, with two notable exceptions: Belgium and West Germany (fig. 7.2). These were the only two nations in which green parties held parliamentary seats at the time of our survey. The Belgian ECOLO/Agalev and the

German Greens use this position to increase legislative action on environmental issues, making the parliament the most responsive agency of government in these two nations. The Danish Folketing receives the highest marks of any parliament, reflecting the progressive stance of its leftist government on recycling, conservation issues, opposition to nuclear power, and other environmental matters. Many Danish packaging and pollution standards are at the forefront of environmental reform in Europe, to the point where some other EC member states claim that these regulations (such as recycling requirements) are an indirect restriction on intra-EC trade (Vogel 1993). British groups also see the House of Lords as slightly more responsive (mean = 2.44) than the House of Commons (mean = 2.11).[7] In most nations, group representatives interlace their "fair" evaluations of national parliaments with positive evaluations for certain parliamentary committees or specific MPs. Overall, European environmentalists find pockets of support within the legislature, even if they view the entire institution as less supportive.

Many European heads of state have paid at least symbolic attention to environmental issues during the 1980s—for instance, Kohl's concern for acid rain and Papandreou's expressed sympathy for Athen's pollution problems—but all of these political figures receive low marks for their handling of environmental problems.[8] In each nation the chief executive receives the lowest score from among these three governmental institutions (often rivaling the abysmal ratings of the ministry of industry). Among Europe's leaders, Ruud Lubbers in the Netherlands and Margaret Thatcher in Britain earned the worst performance ratings. Lubbers has been a foe of the environmental movement since the mid-1970s, when as minister of economics he spearheaded a drive to develop nuclear power in Holland (Hatch 1986). At least one respondent in our survey felt that even the lowest category was too positive for Lubber's performance; in fact, all Dutch groups give him a "poor" mark for his environmental record (mean = 1.00). Margaret Thatcher's ratings were almost equally bad; all but one group assigned her a "poor" score (mean = 1.11). Under Thatcher's administration Britain earned the label of the dirtiest nation of Europe, a reputation that the prime minister sometimes seemed to bear with pride. Official governmental policy was often marked by a failure to recognize the existence of problems such as acid rain or even to conform to international environmental standards. Faced by growing public concern for green issues, Thatcher tried to remake her environmental image in the late 1980s (Franklin 1990; Flynn and Lowe 1992; McCormick 1991), but the legacy of her earlier image remained. In a survey of British executives, a group whose values differ greatly from environmentalists', the *Economist* found that more than 60 percent of executives who rated the environment as an important issue gave Thatcher a "fairly poor" or "very poor" mark for her performance (*Economist,* July 21, 1990, 55). The new government of John Major undoubtedly has a better environmental image, but the negative image of the Conservative Party's commitment to environmental reform likely continues.

The highest-rated executive in our survey is President François Mitterrand of France, in spite of his tempestuous relationship with French environmentalists. After courting their support in the 1981 presidential elections, he reversed his position on the critical issue of nuclear power once in office, a move that earned him disdain among many environmentalists. The 1985 bombing of the Greenpeace ship *Rainbow Warrior* by agents of the French government created further enmity toward Mitterrand, who never fully distanced himself from the incident. Thus Mitterrand's positive rating relative to other European leaders is somewhat surprising, though in leading the pack Mitterrand still receives poor marks (mean = 1.78).

The ratings carry several lessons about the perceptual field of the European environmental groups. The majority of environmental groups are deeply involved in policy making, which leads them to search for potential allies within the institutions of governance. As theorists of New Social Movements claim, this search bears little fruit at an institutional level in most nations. Every European government now has a ministry or department devoted to environmental issues, but these agencies are seldom seen as strong allies of the movement. Similarly, a growing number of politicians are expressing a concern for environmental issues, especially around election time, but this is not translated into perceptions of a responsive parliament. Even more telling, the chief executive represents the culmination of political (and partisan) power within the government, and these officeholders receive consistently poor ratings from environmental groups. Thus, as one moves up a ladder of interest aggregation and presumably political power (MOE, parliament, chief executive), the perceived responsiveness of governmental institutions decreases. Indeed, these patterns are what should be expected of a social movement that is challenging the goals and procedures of the established sociopolitical order.

Business and Labor

The economic interests represented by business and labor remain dominant in most Western European societies, for several reasons. The conflict between these two forces is a major factor in structuring political competition in these societies; the issues raised by these two groups often dominate the political agenda; and the resources available to these groups overshadow those of most other interest groups. Thus, the location of these economic interests in the perceptual field of the environmental movement sharply reflects the political status (and possible partisan orientation) of the movement.

Labor unions often find themselves cross-pressured on matters of environmental reform. Unions traditionally view economic growth as a prerequisite for improving the material standard of living for the working class and as a precondition for social progress. This orientation has frequently led them to oppose environmental regulations that might restrict economic growth, especially those that threaten large infrastructure projects such as nuclear power, highway construction, and airport expansion. As the representative of the specific economic interests of their members, unions generally adopt a protective stance on these issues. Economic recessions during the past two decades have often pushed the unions to emphasize their role as guarantor of these interests over social issues such as environmental reform. For instance, in the tripartite negotiations among government, labor, and business during the recession of the mid-1970s, the German trade unions took an even harsher stance against environmental protection than did the employers (Ewringmann and Zimmerman 1978). In order to build a business-labor alliance in opposition to environmental reform, employers have played on the unions' fears that regulations concerning industrial pollution will lead to layoffs and plant closures (Kazis and Grossman 1982; Siegmann 1985). Contradicting these pressures, unions normally act as a progressive political force in contemporary society, and their members benefit from measures that protect the environment and enhance the general quality of life. In recent years, many individual unions have shown a greater reluctance to equate ecology with the loss of jobs. The ongoing dilemma for the unions is how to balance these two tendencies.

The values and orientations of the business community also reflect a mix of motivations. Industry represents one of the strongest advocates of the dominant social paradigm; surveys of business leaders underscore their continuing commitment to the free market, the supremacy of private property, the acceptance of hierarchical authority patterns, and the creation of wealth as a near moral imperative (Cotgrove 1982, chap. 2; Milbrath 1984). These values and the predictable opposition of most industrialists to environmental regulation lead most scholars to locate the business community firmly within the conflict system of environmental groups. Although the value contrasts between environmentalists and business leaders are indeed sharp, at times the interests of both overlap. There is a potential affinity between business interests and nature conservation groups regarding issues of wildlife protec-

spite of some attempts by business leaders to project greater sensitivity toward green issues, European environmentalists are virtually united in identifying business associations as a major political opponent.[10] The average environmental rating of the major business association in each nation never exceeds 1.50 on a scale of 1.00 to 4.00. Individual businesses or business leaders may support specific green causes, and there were many illustrations of specific alliances in our interviews: a ship-repair firm donated its services to refit one of Greenpeace's ships, a Belgian conservation group was housed in office space donated by a corporation that supported its activities, and several environmental groups worked with specific businesses on individual campaigns. As a group, however, business interests are clearly part of the conflict system of the movement.

Environmentalists give more varied evaluations to labor unions, reflecting the ambivalent relationship between the labor movement and environmental movement.[11] The French unions receive relatively positive scores, because the object of evaluation is the French Democratic Confederation of Labor (Confédération Française Démocratique du Travail, or CFDT), an organization with a long history of sympathy for the environmental movement and New Left causes. The CFDT was one of the first established groups to criticize nuclear power in France and worked jointly with environmentalists in compiling the ecologist slate in the 1978 elections (Nelkin and Pollak 1981). In contrast, the communist-leaning General Confederation of Labor (Confédération Générale du Travail, CGT), which was not included in our survey, is more critical of environmental reform and more likely to approach environmental issues strictly in terms of the economic consequences for their members. Group representatives give the German Labor Federation (DGB) and the Danish Labor Organization (LO) modest approval ratings. The actions of the DGB illustrate the potential for coalition building between Old Left and New Left constituencies. In recent years, the DGB has substantially changed its once-hostile position, especially after the SPD left government and began revising its position on green issues. A major DGB report of 1985 stresses the policy goal of "qualitative growth" and acknowledges that "only environmentally-sound jobs are secure jobs." From a position that environmental reform threatens worker interests, the DGB now maintains that antipollution legislation is a means of job creation. The constituent unions of the DGB are still debating these views, but an expanding network of alliances is building between green organizations and the German labor movement.

Beyond these positive examples, however, the remaining labor federations in Europe receive poor marks from environmentalists. The British Trade Union Congress (TUC) has a mean environmental rating of 1.50, barely above that of the Confederation of British Industry. One British environmental group, the Socialist Environment and Resources Association (SERA), defines its role as the environmental arm of the labor movement. Nevertheless, its representatives claimed that speaking to union leaders was like preaching to heathens (see also Frankland 1990). Because British environmentalists see more support coming from the rank-and-file union members and younger labor officials than from the present union leadership, the prospects for future change are brighter than the present situation. In Belgium, Holland, Italy, and Greece, the perceptions of labor are virtually indistinguishable from those of business (see also Jamison et al. 1990), leading us to believe that the overall prospects for labor-environmental alliances across Europe remain uncertain.

The Movement and the Media

The mass media are especially important actors in the political space of a social movement (Gamson and Wolfsfeld 1993; Schmitt-Beck 1990). For the average citizen, the media provide a window for observing world affairs, allowing individuals to experience firsthand the devastation wrought by an oil spill or to witness threats to endangered species. The media can also be a potent source of popular education on political issues, especially on such new topics as green issues. Beyond these normal information functions, the views and actions of the media are of critical importance to a citizen-based movement. For a large, geographically dispersed popular movement such as environmentalism, the reporting of environmental topics provides a vital information link between the group and actual and potential members. Media coverage furnishes citizens with political cues from the movement and provides some legitimacy for the movement itself. In addition, media coverage can be a key element in the mobilization of popular support. The drama of Greenpeace battling a whaling ship in the small inflatable boats known as zodiacs or of FoE demonstrating against industrial pollution is directly experienced by only a few but is witnessed by many through the media.

The quantity and quality of media coverage of green issues, and the groups themselves, therefore become an important factor in understanding the patterns of environmental action. A sympathetic and interested press can be a vital member of the alliance system among environmentalists. A skeptical and distrusting press can isolate environmental groups and minimize their popular appeal.

There is evidence that journalists are generally sympathetic toward the environmental movement; as an example, small environmental groups or-

FIGURE 7.5

Environmental Performance of the Media, as Rated by Environmentalists

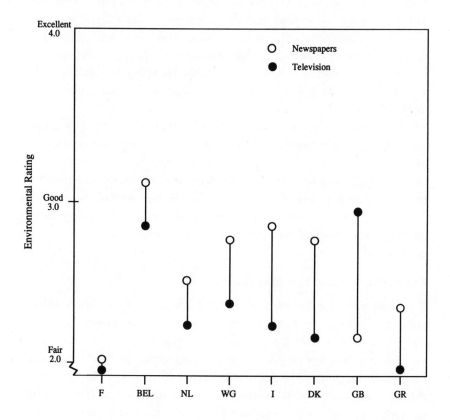

ganized specifically for journalists exist in several nations (Journalists and the Environment in France, and the Green Alliance in Britain). In addition, media coverage of environmental issues has also grown dramatically in the past two decades (e.g., Schoenfeld et al. 1979; Strodthoff et al., 1985).

Our survey found that environmentalists normally perceive the media as supportive of their cause (fig. 7.5). Both newspapers and state television receive relatively positive environmental ratings in most nations.[12] We often heard comments that television was the most important medium for these groups, but newspapers almost always get a more positive rating for their environmental performance. Newspapers can cover a story in more depth than television reporting and are free of the political constraints of state-run television stations. State television receives its best marks in Belgium and

Britain, two nations with an abundance of regularly scheduled programs dealing with environmental matters (Elkington et al. 1988, 197). The poorest marks for television go to France and Greece, where governmental resistance to the movement is seen as extending to government-managed television coverage. Britain is the only nation where television receives higher marks than the print media. Several British environmentalists singled out the positive coverage in elite newspapers, such as the *Guardian* and the *Observer,* and then contrasted that with the poor handling of the same issues in the daily tabloids.

Although there are significant national variations and even variations between specific newspapers and television networks, most environmentalists place the mass media within their alliance network. This support is crucial because of the media's role as a communication and mobilization tool for the environmental groups. Without the media, the movement might not exist. Moreover, this is an active alliance, with media contacts representing one of the frequent activities of environmental groups (see chap. 8).

Networks within the Movement

If reliable allies for environmentalists are scarce among members of the polity, an alternative strategy is to seek support from other challenging groups both inside and outside the movement. Charles Tilly (1978) and others maintain that it is not unusual for challenging groups to band together in confronting the political establishment, even if these groups share no more than an opposition to the political status quo. In addition, the growth of the environmental movement has created a large resource base for action, and if a cooperative spirit prevails, alliances within the movement may create a support network that single groups can draw upon. The peace movement and women's groups are also potential political allies of environmentalists, in that all three of these NSMs share a common critique of the established political order. Many ecology groups were active participants in the peace movement protests of the mid-1980s (Mushaben 1989; Rochon 1988), and groups such as Greenpeace explicitly combine issues of the environment and military policy. Feminism and support for women's issues have been a prominent element in many green parties, linking these two social movements. The potential for alliances among NSM organizations based on their common interests is real.

Although there are many reasons for NSMs to develop alliance networks, these tendencies may be partially counterbalanced by potential competition among them. Environmental interest groups are, to an extent, competing against one another for scarce resources, a finite pool of potential members, and access to the political system. If group leaders act as the self-interested entrepreneurs described in political economy models of interest group politics (that is, they are concerned primarily with expanding their "firm" and only secondarily with goals of political reform), then cooperation among groups might exist only at a superficial level. Any group that is exceptionally successful in generating donations or attracting new members can be seen as depleting the resource pool for the remaining groups.

If we turn to the historical record, there is evidence of both cooperation and competition in the relations among environmental groups (see also chap. 2). For example, during the first mobilization wave of environmentalism (1880–1910), the Royal Society for the Prevention of Cruelty to Animals, formed in 1824, provided financial assistance and administrative support to the Royal Society for the Protection of Birds and other fledgling British conservation groups. The Danish Ladies' Society for the Protection of Animals similarly helped organize the Danish Ornithological Society. This pattern of mutual assistance was even more apparent during the second mobilization wave (1965–1985), because by then the combined resources of the movement were greater. For many ecology groups and their founders, their initial political training and organizational expertise came from their involvement in conservation groups: Friends of the Earth began as an offshoot of the Sierra Club in the United States; Robinwood was a splinter group of Greenpeace in Germany; Actiongroup Straw established the Foundation for Environmental Education in Holland; and the Öko-Institut evolved from the German ecology movement. In some instances the support was more direct: WWF in Holland and Denmark provided the "venture capital" to create the Greenpeace affiliates in each nation and in Germany gave Greenpeace the funds to buy its first large ship. WWF is itself a planned by-product of the International Union for the Conservation of Nature. Another distinctive feature of the second mobilization wave was the creation of multinational environmental groups, which was made possible by an organization in one nation providing the resources to establish affiliates in other nations. Friends of the Earth spread from the United States to Britain and France in the early 1970s, and from there to several other European states. Greenpeace began as a Canadian organization, moved southward to San Francisco, and then crossed the Atlantic to Europe. WWF spread to almost two dozen nations within a single decade.

Interwoven in the historical record, however, is some evidence of competition between environmental organizations. The competition between the RSPB and the Selbourne Society in their approach to environmental reform and their political tactics delayed bird protection legislation in Britain for at least several years (Lowe 1983). Similarly, the National Trust became a central institution in the British conservation movement and its success caused the Society for the Promotion of Nature Reserves, formed several years later, to languish in its shadow (Sheail 1976; Lowe and Goyder 1983, chaps. 8 and 9). Political and resource rivalries sapped the energy of the French environmental movement until the FFSPN was formed in 1968.

Our interviews with group representatives reflected an awareness of both the cooperative and competitive nature of relations within the movement. The majority of spokespersons acknowledged there were rivalries among groups in their nation. In about half the cases these rivalries were based upon the competition for scarce resources, specifically finances and membership. A particularly successful Greenpeace telethon in Holland, for example, generated worries about a dwindling resource pool for other Dutch environmental groups. The groups with large memberships and therefore a richer resource base were looked upon with envy by many smaller groups. In other instances, larger groups felt that the proliferation of highly specialized public interest groups was sapping their membership base.

Another source of rivalry, and actually a more common source, were intergroup differences of a more political nature. Differences based on ideology and strategy nearly equaled resources as a source of competition. Sometimes the differences were fundamental: one Dutch conservation group withdrew from the national environmental federation because it believed the central organization in the federation had supported government actions that undermined its own goals. French and Greek environmental groups were often the most vocal in castigating their colleagues; one French respondent claimed that another group had gone out of existence because of its failures, although the second group continued to function. In many nations we heard reports of how certain groups were too demanding (or not demanding enough) in their policy goals or utilized tactics that were too extreme (or too mild). Environmentalists are sensitive to the various shades of green within the movement and to the tensions that can result from these differences.

Although we identified resource and political rivalries internal to the environmental movement, the predominate impression from our study was a broad pattern of cooperation among environmental interest groups. Our questions about rivalries were often answered by spontaneous comments about the cooperation that occurs between organizations, both in formal and informal ways. This potential for intramovement cooperation is illustrated by another question we asked about whether other environmental groups share the views of one's own organization (the functional equivalent of a set of

TABLE 7.1

Shared Environmental Orientations among Environmental Groups

	Conservation Groups	Ecology Groups	Total
Shared orientation with at least one:			
Conservation group	88%	60%	76%
Mixed group	21	20	23
Ecology group	39	70	48
No shared orientation	9	10	8
Total	157%	160%	155%
(*N*)	(34)	(20)	(62)

Note: Figures are the percentage of each type of group that expressed a shared orientation with one or more other conservation groups, mixed groups, or ecology groups. Totals exceed 100 percent because multiple responses were possible.

questions we asked about political parties; see chap. 9). Nearly all groups note the existence of at least one other environmental organization with similar views; on average more than one other group is mentioned (table 7.1).[13] The most central actors in this environmental network are conservation groups, who are mentioned by three-quarters (76 percent) of respondents. This reflects both the larger absolute number of conservation groups and the commonality of conservation issues throughout the environmental movement (chap. 6). Perhaps the most interesting aspect of these results is the asymmetry of relations between conservation and ecology groups. Nearly all conservation groups see their views shared by at least one other conservation group, while a much smaller proportion see their interests overlapping with ecology organizations. Ecology groups, in contrast, are nearly equally likely to mention another ecology group or a conservation group as sharing their interests. In other words, ecology groups are more likely to see ties to conservation groups than conservation groups are to ecology groups. This repeats the pattern of political networks we observed for career paths (chap. 4). Ecology groups are partially linked to more established interests through their ties to conservation groups.

The cooperative tendencies within the environmental movement are also illustrated by the umbrella groups that bring together environmental organizations in a common forum in almost every nation: CoEnCo and Wildlife Link in Britain; Friluftradet and the Green Alliance in Denmark; the German

Nature Protection Ring and the Working Group for Environmental Questions (Arbeitsgemeinschaft für Umweltfragen) in Germany; the Foundation for Nature and the Environment and the National Platform for the Environment (Landelijk Milieu Overleg, or LMO) in Holland; FFSPN in France; and three separate regional umbrella groups in Belgium (Inter-environment Wallonie, Stichting Leefmilieu, and Raad Leefmilieu te Brussels).[14] In the Belgian example relations between the regional groups are so close that Inter-environment Wallonie and Stichting Leefmilieu occupy the same building, with offices on opposite sides of the central staircase. At the European level, the European Environmental Bureau in Brussels sponsors an annual meeting among national environmental groups, which regularly send more than fifty participants (Lowe and Goyder 1983, chap. 10). In addition, informal working groups dealing with specific problems—such as transportation, nuclear power, or the conservation agreements—are even more common among environmental interest groups. These organizational links provide a vehicle for discussion and the exchange of information, as well as the opportunity to coordinate activities, at least implicitly, by knowing what other groups are doing. Umbrella organizations can also mobilize the efforts of environmental groups and focus their efforts so that they speak with greater force and authority to policy makers.

In addition to these federative structures, environmental groups cooperate extensively on individual projects and specific campaigns. The comments from our interviews and our research on environmental action suggest that the greatest contact and cooperation occur at a personal and informal level. The representative from one large conservation group explained how they were able to use their government contacts to acquire some sensitive policy information for another group; the representative of a Dutch ecology group described a "silent agreement" on the division of effort among certain groups; other respondents talked about how they informally coordinated their policy campaigns with the activities of other groups.

Formal cooperation is also common, however. In overall terms, 60 percent of all groups report that they often work with other environmental groups in these ways, and only 9 percent claim they rarely or never work with others. Such cooperation is equally common among conservation and ecology groups. Indeed, our interviews are filled with descriptions of how groups have cooperated in creating a committee to deal with a specific environmental problem and how they worked together in supporting a specific legislative proposal or even in seeking joint funding for a specific campaign.

In overall terms, the interpersonal and interorganizational networks within the environmental movement are dense. Yet, we also noted distinct national differences in the climate of cooperation among environmental groups. Internal movement networks appear to be the densest in the Netherlands and Denmark; this might be either a cause or a by-product of the successful mobilization of popular support for environmental reform in these nations (see chaps. 3 and 4). Intergroup cooperation also seems fairly common in Britain, where informal links between conservation and ecology groups are frequent. France again stands out for the lack of cooperation displayed by environmental groups, where even formal cooperation was accompanied by informal feelings of alienation and rivalry (see Wilson 1987). A similar pattern of fragmentation seemed to characterize the nations of Mediterranean Europe, but this may be explained by the underdeveloped nature of environmental organizations in these nations rather than by hostility among organizations (see also Diani 1990a).

As a final reflection on these interrelations, we asked group representatives whether they felt the influence of environmental groups was reduced by the many competing organizations. Lowe and Goyder cite one leading British environmentalist who argued that the lack of unity prevented environmentalists from having "real political bite" (1983, 84). Lynton Caldwell has criticized the environmental movement for its fragmentation of effort (1990, 89–97). Similarly, the general thrust of RM theory would suggest that decentralization and informal patterns of interaction would decrease the overall political impact of a social movement. Our respondents do not share these views. Only a minority of group representatives (21 percent) feel the diversity of environmental groups reduces the influence of the movement either "very much" or even "somewhat"; the majority (56 percent) say that it has hardly any effect. Furthermore, these opinions are fairly uniform across conservation and ecology groups. Diversity and pluralism are not seen as sources of weakness.

The Implications of Perceptual Networks

In this chapter we have illustrated how environmentalists view the political world that surrounds them. Our findings reinforce the conclusion that the green movement views itself as challenging the prevailing sociopolitical forces of advanced industrial societies—even if evidence from earlier chapters suggests that the nature of this challenge is more reformist than revolutionary.

Environmentalists of all colors generally consider that most members of the polity are not committed to environmental reform. In some instances environmentalists are confronted by outright opposition, as may happen when dealing with ministries of industry, ministries of agriculture, and business concerns. In many other cases, calls for environmental reform are apparently met by indifference or ambivalence. Even some institutions that are seen as relatively supportive of environmentalism—such as ministries of the environment or the media—receive only modest evaluations. None of the central political actors in these nations falls clearly into the alliance network of environmental interest groups.

Only green/New Left parties are unambiguously viewed as political allies of environmental interest groups, but the radicalism of these parties may actually limit their value in building broader-based political alliances (see chap. 9). The greatest alliance opportunities for environmentalists lie with other social critics—such as the women's movement, peace groups, or antisystem parties—but these actors are also outside the circle of polity members and therefore lack the access to resources shared by groups such as labor, business, and government agencies. In short, the perceptual field of environmental interest groups describes an alliance of challengers aligned against the stronger members of the polity.[15]

The distance between environmental interest groups and the established interests is further illustrated by the variation in environmental ratings as a function of political aggregation. As interests (and political influence) are aggregated to higher organizational levels, there is a drop in support for environmental reform. The perceived support for environmental action is systematically lower for larger political aggregations; some individual MPs are more supportive of environmental legislation than their party overall, some parties are more supportive than parliaments overall, and most parliaments are more supportive than the head of government. The same patterns apply for business interests and the labor movement. Some individual businesses may be sympathetic toward the movement, but the national business association is a clear opponent. Some unions are supportive of environmental legislation, but most national labor federations are perceived negatively.

Our analyses also provide further evidence that the environmental orientations of a group structure its perceptions of the political process. Across Europe as a whole, ecology groups are systematically more critical of the environmental records of major business and political institutions. Moreover, for several institutions the strength of these ideological effects outweighs nationality as a predictor of environmental performance.[16] In other words, the differences between ecology and conservation groups in their evaluations of the MOI or the national business association are greater than the differences in evaluations of these actors across nations. A group's environmental orientations also strongly affect its perceptions of the nation's chief executive and the leading conservative party. As our framework of Ideologically Structured Action would suggest, perceptions of political opportunity structures result from an interaction of objective political conditions and the environmental orientations of a group.

These findings for environmentalists as a whole and ecologists in particular do not preclude the formation of alliances, as RM theorists would suggest, but they do have implications for the types of alliances that might form most easily. Environmentalists' negative perceptions of most members of the polity apparently limit the potential formation of broad political alliances between environmental groups and the established interests of advanced industrial democracies. There are, however, significant variations in the potential for coalition building. Political context may influence alliances, as in Denmark, where a pragmatic environmental movement has been able to build tentative ties to labor, parliament, and the Ministry of the Environment (Jamison et al. 1990). Moreover, the environmental movement is itself expanding the members of the polity to include green and New Left parties that can provide them access to government resources and decision-making processes. Yet there are also important differences in alliance potential within nations as a function of the environmental orientation of specific groups. Ecologists are more critical of most established social and political interests, but conservation groups are relatively more positive in their evaluations of these same actors. Thus another chain of alliance possibilities exists: from ecology group to conservation group, from conservation group to established interests. The environmental movement is not invariably isolated from the polity, but the links will often be weak and indirect rather than based on firm, broad political alliances.

Our findings also carry several implications for the strategies and tactics of environmental interest groups that follow from their negative perceptions of most polity members. If we accept the classification of the overall movement as political outsiders, environmentalists (and especially ecologists) will frequently find themselves dependent on unconventional tactics to upset the prevailing political balance. This might entail dramatic actions meant to pressure public agencies or spectacular events designed to capture media attention and mobilize political support for their cause. Indeed, the one important ally that was missing from these analyses is the broad popular support that environmental reform receives from European publics (see chap. 3). The populist base of environmentalism makes the relationship between these groups and the mass media especially important, because media coverage is vital to developing and mobilizing public support in order to tilt the political balance.

The perceptual field of environmental interest groups also suggests a firm basis for the motto "think globally, act locally." Because political aggregation increases the influence of dominant social interests, a challenging group may be more successful in working with local government or individual businesses than in dealing with prime ministers and national economic associations. Even if business (or labor) as a whole is unsupportive, environmentalists can find allies among individual firms (or unions). At least on a small scale, environmental groups can develop (and have developed) allies that strengthen the movement and expand its influence; or they can build temporary alliances on specific issues. Our interviews contain many examples of this strategy, but one case provides a clear illustration. In developing their program to prevent the dumping of nuclear wastes at sea, the Dutch Greenpeace group was able to assemble an unlikely coalition that included a doctors' association, fishing associations, and the international dockworkers union—all of whom feared the health effects of dumping. Outside of this setting, none of these professional groups is a likely ally of Greenpeace and its general environmental agenda, but in this specific campaign cooperation was possible. Furthermore, with more experience in such efforts, other alliance possibilities may develop.

Chapter Eight

PATTERNS OF

ACTION

Environmental action focuses public and governmental attention on the emerging environmental problems of advanced industrial societies and the structural features of society that produce these problems. Citizens of a small town in Denmark form a society to prevent a stand of trees from being destroyed by developers; residents in Delphi protest a planned aluminum plant that will threaten the ancient temple of Apollo; hundreds of Britons volunteer their time to serve in toad patrols to guide toads safely across busy highways at night; thousands of Dutch citizens work to protect the nation's remaining wetlands. These groups, their members, and their issue interests represent the core of the environmental movement.

Although environmental interest groups are active participants in the political process of most advanced industrial democracies, scholars remain divided on the overall quantity and quality of their participation. The early literature on New Social Movements, and the initial reactions of political elites, often focused on the unconventional political tactics and political isolation of these movements (Brand 1982; Brand et al. 1986; Raschke 1980). The proliferation of citizen action groups, environmental lobbies, and related groups challenged the political establishment by their use of protests, spectacular events, and civil disobedience, as well as by their new issue demands. As shown in chapter 6, the populist orientations of these movements and their status as opposition groups distanced them from established interest groups and political institutions and presumably led them to adopt unconventional methods of political action.

This one-dimensional image of unconventional action by NSMs has been challenged by more recent research that documents the diverse range of tactics actually used by these groups (e.g., Rucht 1990a; Lowe and Goyder 1983; Rochon 1988, chap. 5; Gelb 1989). Resource mobilization theorists also predict that the diverse goals and opportunity structures that environmental groups face lead them to pursue political reform through many routes, ranging from the conventional to the unconventional (Klandermans 1989; Kitschelt 1986).

Beyond this general characterization of the environmental movement, if our framework of Ideologically Structured Action has any value, it should be apparent in the political methods used by various green groups. The mix of political tactics used by environmental groups should reflect their political orientations and their status as organizations challenging the prevailing order—there should be something *new* in the behavior of New Social Movements. Even more directly, we expect that conservation and ecology groups will adopt different political tactics that can be traced back to their underlying political identities and not to the structure of political opportunities. In short, patterns of political action should provide the most direct test of our framework of ISA.

In this chapter we examine two broad topics concerning the political activities of environmental groups. First, we describe the repertoires of action of European environmental interest groups. This allows us to determine the mix of conventional and unconventional methods and the relationship between these various strategies. Second, we examine the factors that influence the choice of activities used by a group. Are action strategies influenced by the institutional structures of national policy making, by the resources of the organization, by ideological orientations, or by other factors? The repertoires of environmental groups, we shall argue, serve as an indicator of both the present status of the movement and its future political prospects.

Action Repertoires

One misconception that we can quickly correct involves the belief that NSMs, including environmental groups, are unconcerned with public policy and uninvolved in the policy process. Certainly there are some environmentalists, especially among the back-to-naturists, who consider environmentalism as a personal creed and a way of life. To them, environmentalism is a source of personal identity and self-transformation rather than a call to lobby the government. For the environmental groups we surveyed, however, efforts to influence national policy on environmental or conservation issues are a prominent part of their activities.[1] One might argue that this is inevitable in that we selected interest groups partially because of their national political prominence. We would point out, however, that the rise of green parties, the policy activities of local environmental groups, and the prominence of green issues on national political agendas all suggest that political activity is a general concern of the movement. The very expansion of the boundaries of politics to include environmental issues was a primary goal of the movement (Offe 1985). Moreover, the ecology groups that make up the vanguard of the contemporary green movement are slightly more likely than conservation groups to say that they are very actively engaged in attempts at policy influence (tau-c = .10). Thus politics is not anathema to environmental groups; it is a central activity if their desires for social and political reform are to be met. A more difficult question to answer, however, is what methods these groups use in attempting to exert policy influence.

We began our inquiry by assuming that organizations need to pursue a variety of methods instead of following a single mode of behavior. The multiple objectives of an organization are one factor that prompts a diversity of action. Dieter Rucht (1982) contrasts activities that are ends-oriented (*Zweck-rationalität* in Weber's terms), those aimed at the expression and reinforcement of movement goals (as represented in Albert Hirschman's [1970] and Frank Parkin's [1972] research), and the type of resource-generating activities emphasized by RM theory. Jeffrey Berry (1977, chap. 8) makes a similar distinction between activities that aim to influence policy makers, to educate or change public opinion, and to mobilize resources and support for the organization (see also Rochon 1988, chap. 5; Schlozman and Tierney 1986; Freeman 1979, 1983). By pursuing a mixture of objectives— for example, pressuring policy makers while simultaneously seeking contributions from potential donors—organizations become active on several fronts.

As groups address their multiple objectives, they find that a variety of strategic choices are available, depending on the goal. Ralph Turner (1970) introduced the now-common distinction among three types of political strategies: persuasion, bargaining, and coercion. For instance, a group might attempt to persuade its target audience (its supporters, opponents, or decision makers) through the logic of its position or the manipulation of symbols (ranging from educational campaigns to attention-drawing media events). Bargaining activities might involve the trading of resources in exchange for policy support, such as financial contributions or electoral support. Coercion might include the use of negative sanctions, such as disrupting business activities or withholding support. Embedded in Turner's analysis is the presumption that because different situations call for different strategies, the effective organization must have a variety of methods at its disposal. In some instances, a public education effort might be sufficient to convince policy makers to address an issue; other cases might require efforts to force policy makers into action.

In overall terms, the political status of the environmental movement should produce a repertoire of action that differs from traditional interest groups, such as labor unions or business associations. Environmental groups should be more likely to emphasize activities aimed at value expression and political mobilization. In part, this is a pragmatic decision; with a widely dispersed individual membership, environmental interest groups must rely on activities that communicate their views to their members, encourage contributions to the group, and keep the organization in the public eye (Berry 1977, chap. 8). Thus, a catchy advertising campaign, a mass demonstration, or a spectacular action should be common elements of their political repertoire. The use of such mobilization tactics should also arise from the populist and antiestablishment values of the movement, especially among ecology groups.

Even when activities are aimed at policy makers, the political status of the group should influence its choice of tactics. As representatives of a movement challenging the political status quo, environmental groups should be less likely to utilize traditional channels (namely, meeting with government officials, consulting with members of parliament, and participating in government commissions) in attempting to influence government actions. This pattern should result both from environmentalists' negative images of governmental actors (see chap. 7) and from the hesitancy of government officials to involve environmental groups in the policy process. Environmentalists' aversion to conventional forms of political persuasion and bargaining should be counterbalanced by a more frequent reliance on protests, demonstrations, and other unconventional forms of action. Social movements generally use such tactics to gain the attention of polity members and to press them to act. In addition to the ends-oriented value of such methods, the use of unconventional political strategies has an expressive value in legitimizing the image of environmental groups as political challengers, a basis of their appeal to their membership.

A brief history of one campaign illustrates the diverse tactics that environmental groups can use and the interactive nature of their activities. The British environmental movement is a long-standing critic of the nuclear power facility at Sellafield (formerly known as Windscale). An event in November 1983 brought this opposition into focus. The wastewater discharge from the plant accidentally included highly radioactive materials that contaminated the nearby Cambrian beaches and spread a several-mile-long radioactive spill across the Irish Sea. Greenpeace acted quickly to draw public attention to the spill and to Sellafield's history of environmental problems. A small group of Greenpeace divers on the ship *Cedarlea* attempted to plug a pipe about a mile offshore that was discharging the Sellafield wastewater. When reports of these activities first surfaced in the media, the government won a High Court injunction to stop Greenpeace, while Sellafield workers fitted a shielding device on the pipe to thwart the divers. For a few days Greenpeace openly defied the injunction, claiming that it was acting to preserve public safety. When it appeared that the management of Sellafield, British Nuclear Fuel, might initiate another suit to attach Greenpeace's assets for violating the injunction, Greenpeace decided to stop its direct action efforts and challenge the Sellafield plant through the courts. The court proceedings provided a public forum for discussing the activities at the plant. Findings of high levels of radioactivity along the Cambrian coast and other evidence presented in court once again highlighted the environmental risks of the Sellafield facility and generated new parliamentary inquiries into its safety. Greenpeace's quiet lobbying activities also mobilized new opposition to Britain's nuclear power program among members of parliament. The High Court fined Greenpeace fifty thousand pounds (approximately seventy thousand dollars) for its actions, which merely provided Greenpeace with a rationale to mount a highly successful direct mail and advertising appeal for contributions. This entire campaign generated new membership and contributions for Greenpeace, new doubts about Britain's nuclear power program, and extensive media coverage of the Sellafield facility and Greenpeace activities. For fifty thousand pounds Greenpeace could not have bought as much publicity as it received through this campaign.

Not all actions in pursuit of environmental reform are as visible and successful as the Sellafield campaign, but Sellafield represents the type of activity for which environmental groups are best known—that is, a public and dramatic event that makes the news. Yet such media coverage does not convey the full range of actions that environmental groups practice.

To describe the activities of environmental interest groups more systematically, we devoted a large portion of our interviews to their political actions, using two approaches. In the early part of the questionnaire, we asked respondents to define the primary policy concerns of their organization over the past year (see chap. 6) and to discuss the methods the organization had used in pursuing these goals (table 8.1). This question details the specific

TABLE 8.1

Group Activities in Pursuit of Primary Policy Goal

Mobilization of public opinion		134%
Inform public	43%	
Media contacts	31	
Spectacular actions	23	
Inform elites	9	
Educational activities in school	8	
Mass demonstrations	6	
Other	14	
Government contacts		111%
Parliament/parliamentary committees	26	
Meet with ministers	17	
Meet with civil servants	14	
Participate in committees	14	
Contact international bodies	9	
Contact local governments	8	
Judicial action	8	
Party contacts	6	
Other	9	
Problem assessment		34%
Research	25	
On-site	3	
Other	6	
Involvement with other groups		24%
Work with environmental groups	11	
Lobby polluters directly	6	
Work with nonenvironmental groups	5	
Other	2	
Direct involvement		15%
Manage lands/historic sites	5	
Organize citizen projects	2	
Other	8	
Other actions		2%
Total		320%
(*N*)		(65)

Note: Figures are percentages of groups that utilized the given activity to address the primary issue of group effort over the previous year.

TABLE 8.2
Methods of Political Action

	Frequency				
	Often	Sometimes	Rarely	Never	Total
Public action					
Contacts with the media	86%	13%	2%	0%	101%
Efforts to mobilize public opinion	72	20	5	3	100
Conventional action					
Informal contacts with civil servants or ministers	53	34	11	1	99
Contacts with MPs/parliamentary committees	53	36	9	2	100
Contacts with local government authorities	45	38	14	3	100
Participation in the work of government commissions and advisory agencies	41	30	20	9	100
Formal meetings with civil servants or ministers	36	52	9	3	100
Contacts with the leaders of political parties	11	44	30	16	101
Challenging actions					
Demonstrations, protests, or other direct actions	25	23	28	23	99
Legal recourse to the courts or other judicial bodies	20	19	30	31	100
Blocking undesired policies by a refusal to cooperate with the government/bureaucracy	13	11	40	37	101

Note: Figures are the percentages for frequency of use of the given activity as a method of influencing politics.

tactics utilized by environmental interest groups, though the response was limited to only one issue (which differs from group to group).

To measure the general pattern of political action used by environmental interest groups, another portion of the questionnaire presented respondents with a list of activities and asked how regularly their organization utilized each (table 8.2).[2] Because we phrased this question in general terms, it should be treated as a measure of a group's predisposition to use various strategies rather than a report of the group's current activities.

One of the most distinctive features of the political repertoire of environmental interest groups is their heavy involvement in expressive and symbolic actions, that is, activities aimed at the public presentation of the movement's goals and the mobilization of popular support. For example, contact with people in the media is the most frequently cited activity; 86 percent of group representatives say their organization is "often" in contact with media figures (table 8.2). Efforts at mobilizing public opinion are second, with nearly three-quarters of the groups (72 percent) claiming to use this approach "often." Even if we focus on specific policy activities (as in table 8.1), the campaigns of environmental groups almost always include efforts to inform and mobilize the public, both as a means of policy influence and as an end in itself.

The second lesson to be drawn from these data involves the mix of conventional and unconventional activities used by environmental interest groups. Despite the antiestablishment image of the environmental movement and its common identification with protest demonstrations and other spectacular actions, the repertoire of environmental interest groups reflects a strikingly conventional mix of political tactics. Most groups lobby government policy makers in some manner. About half of the groups we surveyed (53 percent) claim they often have informal contacts with civil servants or government ministers (only 12 percent claim to engage rarely or never in such activities). Environmentalists report a similarly high level of contacts with members of parliament or parliamentary committees. Interaction with local government agencies and formal meetings with government bureaucrats are also frequent activities for many groups. Likewise, our discussion of the primary policy campaign of each group illustrates these conventional lobbying activities (table 8.1). Meetings with parliamentary committees, government ministers, civil servants, and other direct attempts to persuade and pressure policy makers constitute a central part of the political repertoire of most groups. Although we found that these same groups are openly critical of the environmental record of most governmental institutions, there is still considerable contact between the movement and the government.

In contrast to conventional lobbying tactics, the use of protest actions and other unconventional activities appears remarkably low. A quarter of the groups claim to use protests and demonstrations often, and even fewer rely on judicial means to challenge government actions (table 8.2). Methods aimed at blocking undesired policies—such as resisting government policy or stopping a business from harming the environment—are even less common. Only a minority of groups never use unconventional methods.

What, then, should one make of the unconventional image of the green movement as it compares to this evidence? Like many other parts of this study, patterns of action display a pragmatism and conventionality that is not always apparent in popular accounts of the movement. News reports cover the spectacular protest or the innovative campaign; a breakfast meeting with a minister or a presentation to a local council is less newsworthy and therefore less visible. Our discussions with environmentalists unveiled a sober awareness that policy success requires involvement in the policy-making process, although they are sensitive to the potential problems of working too closely with the government. Ecology groups share these sentiments, and they often keep their regular lobbying activities out of the public eye. Representatives of several organizations stress that the protests and spectacular actions for which their organizations are best known represent but a small portion of their activities—albeit an important one in that such activity focuses public attention on an issue and mobilizes public support for their group. In addition, they emphasize that these campaigns are unlikely to change policy *without* the traditional lobbying activities that they pursued quietly and less visibly.

In spite of their apparent conventionality, environmental groups are probably more likely than traditional economic interest groups to protest and use other direct-action methods and certainly place a greater reliance on efforts to inform and mobilize public opinion.[3] What may be distinctive about environmental interest groups as a whole is the mix of methods they use in the name of environmental reform.

Analyzing Four Modes of Action

The distinctive feature of environmental action might not be the frequency of interaction with other political actors but rather the nature of this interaction. We therefore examined group activities in four areas that exemplify the range of methods available to environmental groups: contact with the media, working with environmental ministries, participation on government commissions, and protest activities.

Media Contact

Environmentalists see the media as one of the few established political actors with a positive record in supporting environmental reform (fig. 7.5). Further, most environmentalists (78 percent) believe that the media are generally sympathetic to their cause. In a world with few allies, environmentalists perceive the media as a distinct exception.

These strong bonds are important because the media directly or indirectly perform essential functions for the environmental movement and for social movements more generally (Schmitt-Beck 1990). Primarily they serve as a communications channel between the groups and their present and potential members. News coverage of events broadens their impact on public opinion and simultaneously provides a tool for mobilizing new support. This communication link is especially important for citizen movements with a geographically and socially dispersed base of sympathizers. Without the media, their actions might go unnoticed except by those who experience them directly, and the movement might not be a political force. Michael Lipsky's (1968, 1151) often-cited analogy about the link between protest and the media can be generalized to many other social movement activities: "Like the tree falling unheard in the forest, there is no protest unless protest is perceived and projected."

The media are also important in creating and reinforcing an identity for the environmental movement. Media coverage helps to define the overall meaning of environmentalism (Gamson 1992). In addition, the political identity of specific environmental groups often rests on their image as projected in the media; this is especially the case for groups with a high public profile, such as campaigning ecology organizations. Media reports are also valuable in locating and mobilizing potential allies outside the movement. For instance, the media campaign of Dutch Greenpeace against dumping nuclear waste at sea was the catalyst for a diverse alliance of social groups supporting new international regulations (see chap. 7).

Finally, the media also can be used to influence policy makers. Compared to direct lobbying techniques that involve bargaining or coercion, the media are an indirect tool of persuasion. Government officials and members of parliament should be sensitive to changes in the political climate, and the political agenda is at least partially defined by issues that attract public interest. Media visibility also gives legitimacy and support to the more direct lobbying activities of environmental groups. When asked about the effectiveness of various tactics of policy influence, most environmentalists (63 percent) rate media contacts as a very effective method, far outstripping other political methods (fig. 8.1).[4]

As we have previously seen, the activity profile of environmental groups underscores the importance they place on media relations. Contact with individuals in the media is the most common activity; only one of the organizations we surveyed had infrequent contact (table 8.2). Efforts to inform the public or to attract media attention are a standard feature of most policy campaigns (table 8.1). Furthermore, two-thirds of all environmental groups report that their contacts with the media have increased over the past five years.

FIGURE 8.1

Effectiveness of Political Tactics in Influencing Environmental Policy, as Rated by Environmentalists

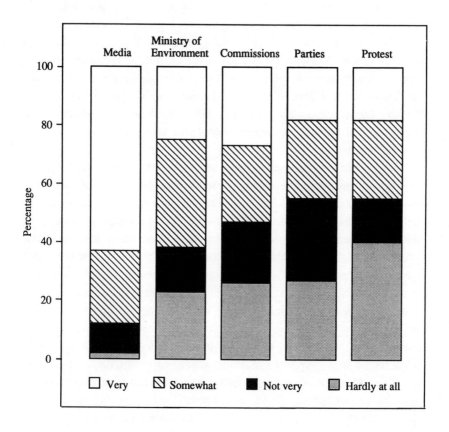

Most interaction between environmental groups and the media involves fairly mundane matters: providing background information to reporters, offering expert opinion on ongoing events or, most important, maintaining personal ties to sympathetic figures in the media. Kielbowicz and Scherer (1986, 85–86) note that becoming a reliable source for news is one strategy that SMOS can use to gain coverage, even though courting the media does not necessarily generate extensive exposure. Rather, these contacts are often an investment in future coverage and an educational offering for the public. For instance, in educating a reporter or assisting her in developing a story about the problems of acid rain, a group is nurturing the consciousness upon which the movement is based, though immediate benefits may be limited.

An environmental group has more visibility (and more control) when it creates or participates in events that are reported by newspapers, television, and other mass media. Such reporting enables a group to focus attention on a critical issue or event while also creating a specific link between the event and their group. As examples, the Royal Society for the Protection of Birds receives nearly as much coverage as Greenpeace in the *Times* of London, the former for its efforts to create bird sanctuaries or protect rare species, the later for its antinuclear protests or high-sea adventures. Whether these events are conventional or unconventional, the primary requirement is that they be newsworthy.

These public events represent a small but crucial part of the media activities of green interest groups by establishing the identity of an organization and transmitting this image to potential supporters. Although many of our respondents extolled the virtues of such events for their group and the movement overall, they also expressed concern about the limitations of this type of coverage. Groups that seek exposure must play by the rules of what is considered "newsworthy"—namely, stories that involve conflict, a dramatic image, or an unconventional occurrence. Thus, the RSPB has to work to create media coverage of a new bird sanctuary, while Greenpeace's protest activities and spectacular actions are often natural candidates for coverage on the evening news. As a social movement becomes known, it runs the risk of being forced into ever more dramatic actions to attract further coverage, which carries a risk of creating a more radicalized image of the movement (Kielbowicz and Scherer 1986). In addition, although this type of coverage generates publicity, it does not encourage an understanding of the broader issues underlying a specific protest. As one of our respondents noted, the media can stimulate the emotion of a movement, but emotion cannot substitute for reason or action.

Another category of media activities involves efforts by environmental groups to produce their own information campaigns. As we noted in chapter 4, most groups distribute a magazine or newsletter to their supporters, through which they provide a more extensive discussion of issues. Furthermore, research by Strodthoff et al. (1985) on the American environmental movement suggests a diffusion of information from these publications into the mainstream press. Environmental groups also utilize the media in direct-appeal campaigns, ranging from advertising for contributions to publishing an open letter to policy makers, endorsed by influential citizens and group supporters.

Our interviews make it clear that environmental groups use the media in remarkably varied ways to pressure politicians, communicate their message, and mobilize resources for the organization. Probably the most successful example of a multimedia campaign is the "Escape from Antwerp" orchestrated by Greenpeace Holland. In 1985 the Dutch Greenpeace ship *Sirius* was attempting to prevent the dumping of toxic waste by Belgian ships into the North Sea. In early May, Belgian naval authorities seized the *Sirius* for interfering with Belgian shipping and escorted it to Antwerp where it was impounded. The seizure alone generated extensive media coverage, but the escapade was just starting. In the dark of a July evening, Greenpeace activists climbed aboard the *Sirius,* disassembled its masts and smokestacks, and cut the chains mooring it to the pier. Then began a slow silent escape to Dutch territory through backwater canals and waterways in northern Belgium. The peaceful liberation of the *Sirius* generated another wave of media attention for Greenpeace as well as for the dumping issue that had prompted their initial action. But the story was not yet complete. Greenpeace arranged to have a camera crew from an independent Dutch television station accompany its crew on the escape mission, so every step in the drama was filmed. Greenpeace then used this film in a telethon to mobilize support for the organization. This high drama captured the public's interest and in the process netted several million guilders (or more than one million dollars) in new contributions for Greenpeace. The escape from Antwerp is an unusual example, but it illustrates the potential that the media have for supporting the efforts of environmental interest groups. A media campaign to build "toad roads" in Britain or to save national architectural treasures from the ravages of air pollution replicate elements of the Antwerp example on a similar, though smaller, scale. Without such interactions with the media, the environmental movement might not exist.

Working with the Ministry

By mid-1986 all of the nations in our survey had a ministry of the environment (MOE) or other agency that dealt specifically with environmental policy or that included environmental policy as one of its prime responsibilities.[5] The MOE is the most obvious access point to government for environmental interest groups: it is the locus for most governmental policy making on environmental issues, the agency responsible for administrating most environmental regulations, and the initial source of most legislation on environmental matters. Moreover, if the environmental movement is to find a receptive ally within government, the ministry is the most likely candidate.

In principle, to successfully implement environmental reform the MOE needs the support of environmental interests just as much as the environmentalists need the ministry's. The potential therefore exists for the movement and the ministry to develop the type of symbiotic relationship that occurs in other policy areas. Indeed, one reason for the creation of a separate environmental ministry is to establish a focal point for policy making and to create an institutional network where the ministry can interact with the relevant policy actors.[6]

At least on the surface, most environmental interest groups seem to recognize the importance of the MOE in the pattern of their political activities. Informal ministerial contacts (primarily with the MOE) are among the most common activities of environmental groups (tables 8.1 and 8.2); formal meetings with ministers and senior civil servants account for a slightly smaller, though still significant, portion of a group's efforts. More than half of our respondents said that the frequency of contact between their organization and the MOE had increased during the five years preceding the interview; less than one-tenth said contact had decreased. Moreover, as we discussed in chapter 7, the MOE is the most positively evaluated government agency in most nations; in several nations, specifically, Denmark and Holland, the MOE receives high marks for its environmental performance.

In discussing ministerial contacts in more detail, however, our interviewees underscore the limits and variations in this relationship. As one form of interaction, the government may take the initiative consulting environmental groups on pending issues. When we asked groups whether the ministry contacts them on matters affecting their interests, some groups claim to be consulted regularly, though these are most often national organizations that focus on governmental relations, such as the Dutch SNM, Britain's CoEnCo, or the Danish DN. The director of DN, for instance, is a former ministry official. These groups often are the official or unofficial governmental contact for the movement and therefore have a special ministerial relationship. One Dutch ecology group, for example, said that, according to a "silent agreement," SNM would perform this role for other Dutch groups. In overall terms, however, most environmental groups lack this consultative status. Nearly half of the groups report that the government rarely or never contacted them about proposed legislation affecting their interests. Even those who are regularly consulted often are openly pessimistic about the efficacy of this consultation. A common response holds that the government discusses policy with them only after the major decisions have already been made; consultation represents a formality rather than a real opportunity for influence.

POLITICAL REPERTOIRES

Conversely, an environmental group may take the initiative and approach the government with a proposal. When we discussed such interaction, a minority of groups (about one-third) again report that they have close ministerial ties and feel there was a sincere effort to respond to their proposals. Most groups, however, see a ministry characterized by indifference or inaction. The majority of our respondents report that they usually receive a sympathetic ear but no real action when they approach the ministry. One German environmentalist illustrated this point by citing the three typical responses of ministry officials to outside proposals: "That is very new"; "We have never done that before"; and "Who is going to pay for it?" This reception is then routinely followed by inaction. We heard this same refrain not only from British and Greek groups, who are generally negative about their respective ministries, but also from certain groups in Denmark and Holland, where the MOE has a more positive image.

If the MOE is the most direct route into the policy process, it nonetheless appears to be relatively circuitous. In most instances, environmental groups are able to gain access to the political process—which can be seen as a considerable advance over the past—but they do not feel that this access is routinely translated into influence. Rather than seeing the ministry as an advocate for their interests, most groups comment on the general lack of interest and concern among ministerial contacts. Consequently, two-fifths of the groups we surveyed say contacts with an MOE are not very effective or hardly effective as a method of influencing policy (see fig. 8.1).

Given this skepticism on the part of environmentalists, one might reasonably ask why they continue to interact so frequently with ministry officials. A partial answer is that without this contact they believe they would exert even less influence than they do presently and that environmental policy would suffer as a result. Even small gains are preferable to large losses. Our discussions also made it clear that not all ministerial contacts are aimed at influence. Environmental groups also maintain ministerial contacts as a source of information and invest considerable effort investigating what legislation is planned or how certain programs are performing. These information networks between environmental groups and the MOE are often dense and based on personal ties. Finally, as with the media, ministerial contact can serve as an investment in networks for future use. Given these various needs, ministerial contact is an important activity for most groups.

The general patterns of interaction we have described can, of course, vary considerably according to group and political context. It is clear from our discussions that both ideology and opportunity structures interact in defining group activities. As the prime illustration of this interaction, paragraph 29 of the German environmental law requires that environmental groups be asked for their opinion on projects that affect the environment as a normal part of German administrative law. Groups that hold this consultative status are regularly contacted by the ministry on pending legislation. It turns out, however, that only traditional conservation groups—BUND, the Federation for Bird Protection, and Schutzgemeinschaft Deutscher Wald— say they are contacted under these provisions, while ecology organizations say they are not included in the inner circle of officially recognized organizations. More generally across Europe, consensual issues generate more frequent and positive interaction than do contentious political issues, which create only limited interaction because of division within the government or between the ministry and the movement. By extension, groups that often raise such contentious issues have more strained relations with the government. Environmental orientations thus structure both the actions of environmental groups and of the ministry.

Participation in Government Commissions

A more intense form of government interaction involves participation in the neocorporatist arrangements that are a common part of European interest group politics (Lehmbruch and Schmitter 1982; Berger 1981; Grant 1985). Neocorporatist politics go beyond normal lobbying activities by uniting government officials and interest groups in settings where policy decisions can be developed among this limited set of actors. Corporatist arrangements, such as government commissions and regulatory bodies, grant a formal political status to the participating interest groups and magnify their potential policy influence.

Our interviews examined environmentalists' participation in such forms of intermediation as a measure of their willingness to engage in one of the most conventional forms of political activity. If environmental groups are simply attempting to gain political stature and maximize their policy influence, then neocorporatist patterns of intermediation should be appealing. The development of neocorporatist arrangements could grant environmental groups a privileged position in the policy process and possibly strengthen the prospects for policy reform.

If, however, environmentalism represents a commitment to participatory politics, direct democracy, and broader goals of social reform, environmentalists should display an aversion to the closed nature of neocorporatist decision making, an aversion caused partially by the policy conflicts that separate environmentalists from the business-labor-government coalition of most neocorporatist systems. For example, data presented in chapter 7 show that environmentalists widely criticize business, labor, and government for their environmental records. In a closed meeting with these three other groups, environmentalists might have little to gain. In addition, the political values of environmentalism should discourage participation in neocorporatist patterns of policy making because of their nondemocratic nature. How environmental groups balance these considerations provides an insight into the values of the movement.

As in other policy domains, European governments have created a host of government committees and study commissions to discuss green issues. Membership on some of the larger and more established committees can represent a potential source of prestige and legitimacy for an environmental group, as well as access to the political process. France and Belgium, for instance, have separate national committees for nature conservation and environmental protection, and most of the larger environmental groups in both nations belong to either or both committees.[7] Many of the larger German environmental organizations belong to the Arbeitsgemeinschaft für Umweltfragen, a consultative body on federal environmental policy. There is an abundance of such permanent bodies in Britain, such as the Nature Conservancy Council and the Countryside Commission. In addition to these standing committees, our respondents described a diverse collection of ad hoc and specialized governmental committees. The prevalence of these bodies seems to vary systematically across nations (being higher in France, Belgium, Germany, and Britain, and lower in Italy, Greece, and Denmark), but enough are available for environmental groups who seek to use this path of influence.

Given these opportunities for action, it is not surprising to find that many environmental groups are often active in governmental committees and commissions (see table 8.2). There also appears to be an upward trend in these activities; 36 percent of the groups reported that their participation in such bodies had increased during the preceding five years, versus only 12 percent who reported a decrease. Furthermore, exactly two-thirds report that they are currently members of at least one such government committee.

Although these data suggest that, despite their antiestablishment rhetoric, most environmental groups willingly participate in government-organized committees—the general tenor of our extended discussions highlighted the difference between the simple quantity of activity and the quality and motivations of action. Many groups that actively participate in government commissions and committees also criticize them on several grounds. Most prominent is a concern that the decision-making dynamics of committees force environmentalists to compromise with other interests. In the words of one Danish respondent: "If we participate in these committees then we have to take the sweet with the sour." Environmentalists frequently cite past experiences that produced mixed results as a reason to avoid committee work. When asked directly, nearly all groups freely acknowledge the potential dangers of cooperating too closely with government in terms of co-optation or a potential loss of support by members who disdain such activities.

Other criticisms focus on the costs of committee work relative to the policy benefits of the activity. Preparing research materials, testifying, drafting proposals, and administrative work are expensive tasks for small under-funded citizen groups, especially when they are pitted against the resources of business, agriculture, or other entrenched interests on the same committee. For instance, several British organizations, including Friends of the Earth, CPRE, and the Campaign for Nuclear Disarmament, participated in the three-year-long Sizewell commission on nuclear power—only to have the commission accept the industry recommendations in its final report. In our interview the FoE representative claimed the group had learned from this experience and would not participate in such commissions again.[8] Other environmental groups had similar, though less extreme stories to tell. Participation in committees that engage in extensive study and research, presumably the more important committees, are an expensive undertaking for any environmental group.

Even if environmentalists are able to influence a committee's recommendations, translating these recommendations into policy remains another obstacle to be surmounted. The likelihood of policy impact is almost inversely related to the extensiveness of the committee's recommendations; the more broad-reaching the proposal, the more likely the opposition from other interests such as business, agriculture, or labor. Taking these caveats into account, it is not surprising to find that environmental groups are almost evenly divided in rating the effectiveness of government committees as a means of policy influence (see fig. 8.1), ranking the efficacy of committee action slightly below ministerial contact.

Our discussions about government committees also illustrate the role of environmental orientations in structuring this relationship. Doubts about the effectiveness of government commissions as a tool of environmental reform were scattered throughout our interviews but were especially noticeable among ecology groups. Ecologists are uncomfortable with the bureaucratic style of these bodies and the resource demands of committee work. In addition they see tension between their broad political goals and the moderate, consensual recommendations that normally flow from such committees. Conservation groups, in contrast, are often more closely integrated into governmental structures and more at ease working inside the system. As an example, the Dutch Greenpeace representative said that her organization had participated in such governmental committees only once in its history; committee meetings were boring and accomplished little. She then volunteered: "This is what the Foundation for Nature and the Environment is paid by the government to do—it is their job and we leave them to it." As partial validation of this division of effort, the SNM representative boasts that the group sits on fifty to sixty government committees and commissions (see also Moltke and Visser 1982, 63–64). This imbalance appears across our sample; 80 percent of conservation groups say they are a member of at least one government body, versus 37 percent of ecology groups.

Like ministerial contact, some groups take part in government commissions and committees as a means of influencing policy, but many other groups see participation as the price they must pay for wanting to be involved in the policy process. Even if the impact of these bodies is limited, some groups participate because of the supplementary benefits, such as political stature or access to information and government officials. Nevertheless, neocorporatist structures may have limited value for many environmental groups. Participation in government bodies often calls for large expenditures of resources. For ecologists especially, participation in government committees may be an ineffective political tactic: close contact with the government may force them to moderate their political demands and may undermine their legitimacy as challenging groups. An even more fundamental question concerns the issue of whether participation in government bodies undermines the movement itself. First, by seeming to accommodate environmentalists, the government decreases citizen pressure for reform. Second, by drawing together diverse interests within a single committee, the government is able to thwart or delay policy action. In other words, such neocorporatist bodies may prove a greater benefit to the government in moderating potential opposition and institutionalizing political conflict than to environmental groups in their efforts to effect policy reform. Participation in these committees thus represents the tension between environmentalists' adherence to their broad political goals and their desire to influence specific environmental policies.

Protest Activities

Environmental groups that seem ambivalent toward the conventional channels mentioned above may be drawn to the use of protests, spectacular actions, and other forms of unconventional political participation. Protest behavior has become a planned and organized activity in contemporary democracies, and environmental groups often appear at the vanguard of this development (Nelkin and Pollak 1981; Brand et al. 1986). The unconventional political tactics of green groups are sometimes traced to their policy differences with the business-labor-government coalition in most European societies. The reliance on such direct-action techniques is also linked to the participatory values of NSMs that stress methods of direct democracy. Moreover, many ecology groups trace their beginnings back to specific protests or political events, and this political style lives on within these organizations. Greenpeace began with their crusade against the Amchitka nuclear test; FoE-UK dramatically announced its creation with an attention-grabbing campaign against Schweppes no-return bottles; Robinwood first gained prominence for its antinuclear actions; NOAH began with a virtual assault against academic experts on the environment (see, e.g., Brown and May 1989; Lowe and Goyder 1983, chap. 7; Jamison et al. 1990, 66). Even moderate conservation groups have adopted protest activities to publicize their cause and their organization. To be green is to be unconventional.

Because the environmental movement is closely identified with the use of protest and other forms of direct action, it is surprising to see how infrequently environmental groups actually resort to these tactics (table 8.2). Although our sample includes many campaigning groups, the frequency of challenging actions is fairly low; 25 percent engage often in demonstrations and protests, 20 percent take legal action through the courts, and only 13 percent regularly use obstructionist tactics. Furthermore, although other forms of political involvement—including conventional actions—show a trend of increasing use, most groups report a relatively flat trend in protest activities: 68 percent said their level of activity had remained the same during the preceding five years, 15 percent were more active, and 16 percent were less.

These modest levels of protest activity temper the unconventional image of the environmental movement presented by NSM theorists. Furthermore, expressions of a declining reliance on unconventional tactics emerged throughout our discussion of protest activities. For instance, the director of one ecology group spoke of "demonstration tiredness," explaining that demonstrations were passé and demonstrators more difficult to turn out. Large-scale protests no longer attract the public's attention as they did a decade ago. In short, the rhetoric of unconventionality is not matched by reality.

We also should realize, however, that unconventional political activities differs from traditional lobbying techniques. First, protest activities are generally not aimed at directly influencing policy makers.[9] Unlike labor organizations, environmental groups cannot use strikes and economic power to coerce the government; nor can they bargain through the threat of a strike. Only a few of the most radical environmental groups attempt to use violence as a basis of coercion (see, e.g., Manes 1990). Most discussions of violence in our interviews carried a negative tone, with respondents noting that if protests became violent, the media and the public would turn against them. Because of the popular support for environmental reform, attempts at coercive violence may be counterproductive. Instead, environmental groups use protests and spectacular actions primarily as expressive and resource-generating activities in order to focus media and public attention on an issue. In the realm of such actions, environmental groups use a creative array of tactics to advertise their cause: one group poured fluorescent dyes into the discharge pipe from a large corporation that was pouring effluents directly into the sea; another group built a "human bridge" across the Rhine to dramatize pollution in the river; a conservation group built a "wall of lead" outside the Danish parliament to demonstrate how hunters' lead pellets are poisoning wildlife; another group organized a large public rally against acid rain; and one group released hundreds of balloons outside a British power station to demonstrate the spread of airborne pollutants. These actions enable environmental groups to bring their issues to the political agenda and perhaps to mobilize more supporters in pursuit of their cause.

Second, protests and spectacular tactics differ from conventional ones in their potential application. The impact of lobbying techniques probably increases with use, as political contacts are made and developed. In contrast, political observers describe direct action as a method that loses impact with use and so must be applied selectively. Weekly protests will dilute the unconventionality of such actions, and thus the potential for attracting media coverage and public attention. Too frequent efforts to mobilize supporters and organize large-scale protests may also exhaust the energy of the movement and its supporters.[10] Unconventional tactics are therefore most effective when they are infrequent (and novel).

The logic of intermixing protest and conventional activities was starkly illustrated in our interview with one well-known campaigning group. After elaborating on the group's protest activities as part of their prime policy activity of the previous year (as presented in table 8.1), the respondent nevertheless rated protest activities as an infrequent activity for the group. Noting his own apparent contradiction, he explained that the group consciously planned its dramatic actions to generate media attention for a specific issue goal: "Protest has no value in itself; it only captures public attention and other actions are necessary to carry through policy change." Once they brought public attention to the issue, the group then worked behind the scenes with sympathetic government officials and MPs to press for policy reform. This tale was often repeated in our interviews.

Predicting Action Patterns

As we have seen, the environmental movement overall has developed a varied repertoire of political tactics, and the most successful groups are probably those that use a combination of complementary strategies. The dividing line that once may have separated conventional and unconventional action is no longer clearly drawn. Many traditional conservation groups find themselves using unconventional tactics to draw publicity to their cause, even if these tactics still appear modest in comparison to the campaigns of ecology groups. Conversely, many ecology groups find themselves drawn into conventional methods of lobbying to achieve their policy goals. Most environmental groups now realize the need for political action appropriate to the issue at hand. As Dieter Rucht (1990a) has observed, the distinctive feature of NSMs may be their reliance on both the conventional and unconventional methods of interest-group politics.

Having a diversity of tactics does not mean that a group uses each method available to it or uses the methods they do choose with the same regularity. Each group develops its own political style, choosing a pattern of action consistent with its goals, its resources, and the nature of the group. For instance, the leaders of several large, well-established conservation groups told us that they could not use direct action methods without alienating their members. In contrast, many ecology groups seem to live for direct action. Some groups have the resources to maintain a regular presence in the policy process, attending committee meetings and issuing policy reports; others lack the resources to engage in policy making at this level even if they wanted to.

In this section we examine the factors that influence the political activities of environmental groups. The set of relevant factors is partially dependent on the level at which we discuss political action. If we focus on the *tactics* of a single campaign, which might include any of the activities presented in table 8.1, the choice of methods becomes closely linked to the specifics of the situation and must be conditioned by calculations of which method will be most effective. Such factors as the nature of the issue, the target of the action, lessons drawn from past campaigns, and the like affect the final decision (see, e.g., Turner 1970; Freeman 1979, 1983; Berry 1977, chap. 8). Our study, however, is primarily interested in the overall patterns of action, what we have termed *strategies* of action, which guide the general behavior of environmental groups. In other words, why are environmental interest groups predisposed toward certain modes of action and hesitant to engage in others? This emphasis on strategies focuses attention on the patterns of action of environmental groups, and from this we abstract conclusions about the political tendencies of the green movement. In addition, claims about the political distinctiveness of NSMs are based on assumptions concerning the strategies that typify these movements (Rucht 1990a; Brand 1982; Brand et al. 1986).

Resources. Social movement theory often emphasizes resources as a factor influencing the action strategies of social movement organizations. In absolute terms, the amount of organizational resources should be related to the overall political activity of a group (Schlozman and Tierney 1986; Walker 1991, chap. 6; Turner 1970; Berry 1977). Resources should be especially important for activities that place heavy demands upon a group, such as participating in commissions or closely monitoring the legislative process. Even organizing a protest or mass demonstration requires a significant investment of personnel and other resources if it is to be effective. As Stephen Barkan (1979) has noted, the well-endowed organization has the autonomy to choose its methods and to pursue a variety of activities simultaneously.

The nature and source of a group's resources may also significantly influence its pattern of action. In chapter 4 we determined that environmental groups differ in their resource base. Some place great reliance on governmental and foundation support, which should be related to a more conventional style of political action. Other groups draw support primarily from membership contributions, which then encourages a style of political action consistent with the political norms of their members.

Organizational Characteristics. A second factor related to choice of strategies is the structure of an organization. Mass-membership groups, for instance, should emphasize activities aimed at mobilizing resources and popular support for the group. Research institutes and environmental federations are more likely to pursue conventional and result-oriented political activities. In addition, we would expect the organizational structure of an environmental group to be mirrored in its political tactics. Groups that emphasize participatory politics and decentralized structures are more likely to adopt participatory direct-action methods, whereas centralized and bureaucratic groups may demonstrate a greater willingness to engage in conventional forms of political lobbying (Walker 1991, chap. 6).

Opportunity Structures. In addition to internal characteristics, resource mobilization research has emphasized the role of institutional context in social movement behavior (McAdam 1983; Klandermans et al. 1988; Jenkins and Klandermans 1994). RM theorists, because they view SMOs as rational actors responding to opportunities as they arise, assume that the structure of those opportunities will influence the strategies chosen. For instance, if neocorporatist activities represent a real opportunity for influence, the sensible organization will utilize these methods; if conventional lobbying is more effective, this mode will be preferred. The importance of such opportunity structures has been an important theme in research on NSMS. In a provocative article, Herbert Kitschelt (1986) mapped the action modes of antinuclear power groups in four nations as a function of the political structure of the respective nation; he then extended this analysis of the political environment in his work on green parties in Germany and Belgium (Kitschelt 1989; Kitschelt and Hellemans 1990). Similarly, in her comparative analysis of the feminist movement, Joyce Gelb (1989) emphasizes opportunity structure as a major influence on national women's movements. Women's groups in America appear most conventional in their style and tactics, while the British have remained unconventional. Researchers have also examined the importance of these structural factors for the peace movement (Klandermans 1990; Kriesi 1985; Rochon 1988), American public interest groups (Berry 1977; Schlozman and Tierney 1986), and other NSMS (Wilson 1990). Indeed, the logic that political organizations should respond to the structure of political competition seems to be an obvious fact.

Ideology. A key element missing from most prior research on political strategy has been the ideological identity of an organization, which should have a direct impact on political activities. To use the terminology of David Snow and Robert Benford (1988, 201), the identity of a social movement organization leads to *prognostic framing* of what strategies, tactics, and targets are appropriate to the group. Because conservation and ecological orientations reflect a group's relationship to the prevailing sociopolitical institutions, as well as its beliefs about democratic politics, its identity should predispose it toward modes of action that are consistent with its self-image. Ideology thus frames a group's perceptions of which tactics are appropriate and likely to be effective. The highly regarded head of a British conservation group told us that because the basis of power in Britain is the bureaucracy, knowledgeable groups focus their efforts on civil service contacts; two days later the representative of a British ecology group maintained that the only way to change policy was through mobilizing public pressure from outside the system.

The political activities of a group are also important as a vehicle for communicating its identity to its present and potential supporters. The representative of a large national bird association commented that his group would never engage in dramatic protests because such actions would alienate many of its conservative members; just as obviously, the spectacular actions of Greenpeace are its main recruiting tool. In short, the style of an environmental group is both a means of political influence and an expression of identity.

As discussed in chapter 4, the environmental orientation of a group also indirectly affects the choice of action strategies by influencing both the structure of environmental groups and their resource base. Conservation groups tend to be dependent on government and foundation support, which might be jeopardized should they adopt unconventional tactics. In contrast, the resource base of ecology groups might reinforce (or facilitate) a challenging, unconventional form of political action.

The ideological identity of an environmental group can interact with opportunity structures. As we have previously observed, opportunities for cooperation with government agencies or the potential for using other action strategies is linked to the environmental ideology of a group. The bird society in a nation, for example, faces a different set of political opportunities than does a radical ecology group. In this vein, Dieter Rucht (1990b) has criticized Kitschelt's emphasis on national opportunity structures, claiming that Kitschelt ignores both the role of ideology in explaining national patterns and the temporal variations in political strategies used by antinuclear groups, even while opportunity structures remain constant. Edward Walsh (1989) similarly found that the tactics of American groups protesting Three Mile Island were influenced by the broader political values and ideology of the respective group. Indeed, our interviews document large intranation variations in political strategies that cannot be explained by national opportunity structures.

Figure 8.2 thus summarizes the factors that should be most influential in shaping the general strategies of action pursued by European environmental interest groups. The environmental orientation of a group influences its choice of methods directly as well as indirectly through the patterning of group resources and organizational structure. Opportunity structure represents a contextual influence that partially transcends the characteristics of specific groups; but it, too, is partially a function of the environmental orientation of a group. Our goal is to test the components of this model of action.

FIGURE 8.2
Factors Affecting the Choice of Political Strategies

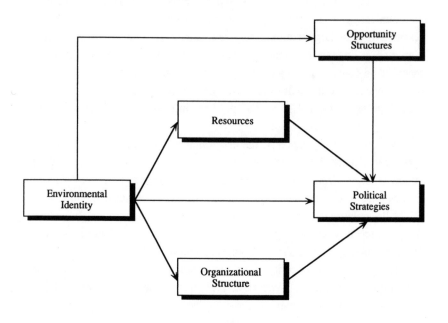

We began our analysis by measuring four types of political action: media contact, cooperation with ministries, participation in commissions and committees, and use of unconventional actions.[11] We shall now look more specifically at the relationship between organizational resources and patterns of action (table 8.3).

As we expected, groups that are large, well funded, and well staffed are able to choose from a broad array of political activities; a large resource base enables them to engage in frequent ministerial contacts, to participate in government commissions, and to be involved in media activities. The resource relationship seems to break down, however, in the case of protest activities. There is a weak (though statistically insignificant) tendency for protest to be less common among groups that are well staffed and well financed; only groups with a large membership display a statistically significant positive relationship with unconventional political activities. To an extent, these data support Michael Lipsky's (1968) dictum that protest is the politics of the powerless, and organizations with abundant resources are more likely to work through conventional channels.

TABLE 8.3

Correlation of Organizational Resources and Political Action

	Type of Activity			
	Ministry Contact	Commission Work	Media Contact	Protest
Size of professional staff	.29*	.09	.01	.09
Adequate size staff	.29*	.07	−.03	−.08
Size of membership	.17*	−.07	.09	.12*
Annual budget	.19*	.33*	.11	−.06
Ranking of income sources				
Membership dues	−.11	−.18*	.07	.29*
Individual gifts	−.06	−.12*	−.05	.11
Foundation support	.07	.03	.03	−.09
Government grants	.07	.31*	−.03	−.06

Note: Figures are tau-c correlation coefficients. Coefficients marked by an asterisk are statistically significant at the .10 level. See footnote 11, this chapter, for information on the construction of each activity dimension.

Even more intriguing than absolute resource levels is the relationship between a group's source of income and its pattern of activity. As we suggested earlier, the nature of a group's funding reflects both its political identity and its position in the organizational field of European politics. Organizations relying on individual membership dues and donations are likely to place a higher priority on expressive tactics that focus public attention on their issues. Groups relying on the central government or foundations for a significant portion of their income would reasonably be expected to lean toward conventional methods of political lobbying. This contrast is clearest when comparing participation in government committees and participation in protest politics. The importance of membership dues as an income source is negatively related (tau-c $= -.18$) to participation on government committees and positively related (.29) to unconventional political action. Conversely, the importance of government grants is positively linked to committee activity (.31) but negatively related to protest activity ($-.06$). Ministerial relations mirror the pattern of committee participation, though with weaker effects. In

TABLE 8.4

Correlation of Organizational Characteristics and Political Action

	Type of Activity			
	Ministry Contact	Commission Work	Media Contact	Protest
Age of organization	.17*	.33*	−.03	−.34*
Organizational structure				
Mass-membership group	−.07	−.25*	−.02	.14*
Leadership control	.08	.21*	.04	−.07
Membership input	−.08	−.01	.01	.18*

Note: Figures are tau-c correlation coefficients. Coefficients marked by an asterisk are statistically significant at the .10 level.

overall terms, financial incentives appear to reflect and reinforce the choice of political repertoires used by environmental interest groups.

Organizational characteristics are also relevant in analyzing the political activities of environmental groups (table 8.4). The first row in the table shows that older environmental groups are more likely than newly formed ones to participate in conventional forms of action, such as ministerial contact and government commissions, and are less likely to engage in protest activities. These relationships provide at least partial support for a life-cycle theory of the gradual institutionalization and incorporation of interest groups as they age and become accepted by the political establishment. The RSPB, for instance, was considered an activist protest group during its founding period because of its letter writing campaigns, media exposés on bird hunting, and public protests against the feather trade (Sheail 1976). As the organization and its goals became increasingly accepted by society, however, the group evolved into a conventional emphasis group. In addition to this life-cycle pattern, these relationships also likely reflect generational changes in the composition of the environmental movement. Conservation groups were formed primarily during the first wave of environmental mobilization, and antiestablishment ecology groups are a newer element of the movement (see chap. 4).

The bottom panel in table 8.4 describes the link between the structure of an organization and its patterns of action. As implied above, mass-membership groups are less likely to engage in government committees and

TABLE 8.5

Correlation of Opportunity Structures and Political Action

	Type of Activity			
	Ministry Contact	Commission Work	Media Contact	Protest
System characteristics				
Neocorporatism	.15*	.24*	.19*	.26*
Leftist government	−.24*	−.01	−.16*	−.08
Viable New Left party	.00	.15	.04	.21*
Openness of system	.08	.10	.11*	.01
Perceptions of government				
Ministry of environment	.08*	.19*	.03	−.21*
Parliament	−.12*	.05	−.05	−.17*
Chief executive	−.04	−.10	−.12*	−.19*

Note: Figures are tau-c correlation coefficients. Coefficients marked by an asterisk are statistically significant at the .10 level.

to consult formally with ministries; these are more often the domain of national federations, research institutes, and elite organizations. In addition, mass-membership groups are more likely to engage in unconventional forms of political action. There is also a weak tendency for the internal distribution of power within a group to influence patterns of action. Groups that are controlled by the leadership or have limited means for the membership to influence group decisions are more likely to engage in conventional political tactics, such as government committees and ministerial activity. Decentralized and more populist groups display a weak corresponding tendency to pursue unconventional methods of political action.[12] We expect, however, that these structural differences are largely a by-product of the relationship between ecologist or conservationist orientations and group structures.

We also examined the correlation between national opportunity structures and the activities of environmental groups (table 8.5).[13] In neocorporatist systems environmental groups are likely to engage in ministerial contacts (tau-c = .15) and to participate in governmental commissions (.24), but they are also likely to have active contact with the national media (.19) *and* to participate in protest activities (.26). This pattern suggests that neocorporatism, rather than defining an opportunity structure that encourages certain

TABLE 8.6
Correlation of Organizational Values and Political Action

	Type of Activity			
	Ministry Contact	Commission Work	Media Contact	Protest
Orientation of group				
Ecological orientation	−.21*	−.36*	.10*	.42*
Leftist ideology	−.30*	−.15*	.16*	.38*
Orientation of respondent				
Satisfied with democracy	.12*	.30*	.07	−.17*
Favor radical social change	−.21*	−.38*	.05	.27*

Note: Figures are tau-c correlation coefficients. Coefficients marked by an asterisk are statistically significant at the .10 level.

political strategies and discourages others, might be spuriously discriminating between nations with a very active environmental movement and nations with less active movements.[14] Other factors, such as leftist control of government, the presence of a viable New Left party in parliament, or governmental openness to interest groups, have weaker and inconsistent correlations with activity levels. In overall terms, national political structures generally exert only a modest impact on the activities of environmental groups, and the pattern is often inconsistent with what might be expected.

We also looked at opportunity structures in perceptual terms, determining whether environmentalists' evaluations of the performance of government institutions are related to patterns of action (table 8.5; see also chap. 7). Positive evaluations of the ministry of the environment are linked to a greater use of conventional political tactics, especially participation on government-sponsored committees.[15] Conversely, groups that are critical of governmental institutions are more likely to participate in unconventional methods. In other words, environmental groups' perceptions of their organizational field are related to the political strategies they adopt. Again, however, even perceptual measures of government agencies have only a modest influence on a group's choice of strategies.

The final factor to consider involves the role of organizational values in guiding political strategies (table 8.6). Ecology groups are significantly less likely to meet formally with ministry officials (tau-c $= -.21$) and to partic-ipate on government committees $(-.36)$; they also are more likely to engage in unconventional tactics $(.42)$ and to utilize the media $(.10)$. The Left/Right orientation of a group mirrors this same pattern. If we examine the other activities included in table 8.2 (data not shown), we find that environmental orientations are strongly related to two modes of political behavior. Conser-vation groups rely primarily on conventional political action, such as min-isterial contact, government commissions, working with local government, or dealing with MPs and parliamentary committees; they sometimes use protest, but it is an infrequent activity and normally modest in scale. In contrast, ecology groups are more likely to use unconventional tactics, while also actively engaging in conventional lobbying methods.

In examining whether the political values are related to the action pat-terns, we found that environmentalists who are satisfied with democracy are more likely to represent organizations that are oriented toward ministerial interaction (tau-c $= .12$) and participation on government committees $(.30)$ and are less likely to engage in protest activities $(-.17)$. Similarly, those environmentalists who favor radical social change represent groups that ac-tively use unconventional political methods and are less likely to use tradi-tional methods of political lobbying. These data reaffirm the pattern we observed in chapter 5, in which the personal values of environmental elites overlap with the values of their organization.

Our results suggest that a variety of factors at least partially condition the political strategies of environmental interest groups. To provide an esti-mate of the relative importance of these various predictors, we computed multiple regression models for each political activity (table 8.7). The large number of potential factors and the relatively small sample size means that we could not incorporate all these measures into a single regression model. Moreover, many of these variables are strongly interrelated, so the multicol-linearity of measures would produce misleading results. For instance, various measures of organizational resources are highly correlated, such as the size of a group's membership and its annual budget $(r = .88)$; it is therefore difficult to separate statistically their distinct effects if both are entered into a single model. To address these problems, we used a theoretically driven method of model specification. Given our theoretical assumption that envi-ronmental orientations are the primary force behind patterns of political action, we first entered this variable into each regression model if it was statistically significant in the bivariate relationship. (Only media contacts

TABLE 8.7

A Multivariate Analysis of Political Action

	Regression Coefficient	Pearson Correlation
Formal contact with ministry		
Adequate staff	.33	.37
Ecological orientation	−.20	−.26
Multiple R	.33	
Membership in government commissions		
Size of mass membership	.45	.40
Importance of government grants	.45	.43
Ecological orientation	−.17	−.41
Multiple R	.67	
Media contact		
Satisfaction with democracy	.12	.12
Multiple R	.12	
Protest activity		
Ecological orientation	.52	.52
Membership influence	.25	.25
Multiple R	.57	

Note: Figures are the results of regression analyses of each political activity. See the text for the description of the methods used for determining these models. Each of the parameters in this table is significant at the .10 level, except for "satisfaction with democracy" in the media model.

displayed no significant correlation). We then identified a list of other potential predictors from the preceding analyses.[16] A forward selection procedure (with a .10 significance level) was used to include additional variables in the model that had a significant effect in predicting activity patterns.

The results of these regression models underscore the importance of environmental orientations in structuring the political actions of environmental interest groups. Environmental orientation is the strongest predictor of participation in protest activities ($\beta = .52$) from among the predictors included in our list.[17] In addition, ecology groups are significantly less likely to have formal contact with ministerial officials $(-.20)$ and to participate in government commissions $(-.17)$. In addition to these direct effects, environmental orientations also have significant indirect effects through their influence on the other predictors in each model. In very clear terms, ideology does structure political action.

The regression models also highlight the importance of organizational resources in influencing political strategies. The adequacy of staff support, a subjective measure made by the interviewer, is an important predictor of ministerial contact, and the size of the membership base is strongly related to participation in government commissions. In addition, government funding is a strong predictor of participation in government commissions $(\beta = .45)$. Like animals in nature, one finds environmental groups where they feed. Because of the statistical overlap (multicollinearity) between these various resource measures, the specific variable identified in each model is less significant than the fact that a resource variable appears in both models of conventional political action.

Two sets of predictors do not appear to carry independent weight in these multivariate analyses: first, none of the measures of national opportunity structures emerged in any of these models. We would argue that despite the attention given to these factors in prior research, one should not expect broad systemic structures to have strong effects on the behavior of individual environmental groups such as those we surveyed. Greenpeace will act like Greenpeace, whether it is in Britain or Italy, just as the political actions of the British and Italian bird societies have more in common with each other than with the Greenpeace affiliates in their respective nations. Any systemic differences between nations are therefore outweighed by the variation across groups within nations. The organizational characteristics of environmental groups also seem to have only a weak impact on action strategies. Membership influence increases the propensity to engage in protest activities, but no other feature of organizational structure appears in the other models.

The one anomalous model is media contact, where no single predictor is significantly related to contact. The media are apparently so important to social movement organizations that they are used in almost equal frequency by all environmental groups, virtually independent of the ideological and resource factors that influence other types of environmental action. Yet even in this case, the strongest single predictor is the political orientation of a group's spokesperson (satisfaction with democracy) rather than the group's resources or structure.

Conventionality and Unconventionality

Evidence presented in this chapter tempers assumptions in prior research literature concerning the political isolation and unconventional style of NSMs. The action repertoires of environmental interest groups appear more conventional than initially assumed. Although a dramatic event—such as the protests following the Chernobyl disaster, the exploits of Greenpeace campaigners, or the hanging of banners from monuments in Rome—might easily capture our attention, more conventional forms of lobbying activities are more common among environmental interest groups. In absolute terms, participation in government committees, meetings with government officials, and other regular lobbying activities outweigh involvement in protests, demonstrations, and other unconventional forms of political action.

Involvement with government institutions and the use of conventional political activities do not mean, however, that the political repertoire of environmental interest groups is identical to established interest groups. We find, for example, that the most common activities of environmental groups are aimed at creating and mobilizing public support for environmental reform, such as working with the media or other public information efforts. This emphasis on the public base is one factor that separates environmentalism from such traditional economic interests as business or labor. In addition, although conventional activities outweigh unconventional ones, their combined use is a distinguishing feature of the environmental movement. Many environmental groups consider protest activities as a standard part of their political repertoire, even if they go beyond this method in pursuit of policy influence.

We view this diversity of political methods as a source of strength for the environmental movement. Rather than being constrained to work only through conventional channels or being so isolated from routine politics that the only option is to take to the streets, the environmental movement can legitimately work in both domains. This situation was illustrated during one of our interviews. The director of one established conservation group took time during the interview to answer telephone calls from several socially prominent individuals who were willing to have their names added to a letter calling for a specific legislative proposal that the group was publishing the following day in the national newspapers. He pointed out that these same individuals would refuse to lend their names to an activity organized by any of the leading ecology organizations. In fact, however, FoE was working with its own supporters to back the same proposal. One lesson that emerges from this example is that a variety of methods enables the environmental movement to mobilize a broader range of supporters than could be attracted by any one group or any one campaign.

The above example illustrates another lesson drawn from our findings: the range of tactics described for the environmental movement in general is not necessarily available to all groups. The heads of several conservation groups openly praised the unconventional political tactics of some ecology groups but were equally quick to note that such tactics were not feasible for their organization. The nature of their issues and the political tendencies of their supporters led them toward conventional channels of political influence. Conversely, ecology groups and their members seem to live for the excitement and drama generated by spectacular actions and other unconventional efforts, though they also have the option of conventional processes.

Often the attempt of social science research to find regularity in patterns of action results in an emphasis on one mode of behavior as optimal for influencing policy (e.g., pluralist or neocorporatist theory). Because of the assorted issues involved in conservation and environmental policy and the present fragmentation of policy making on these issues, the ideal pattern of action for the environmental movement may be the diversity of methods we have observed. Through this approach groups can enlarge their pool of potential supporters and exert multiple pressures on the government, and perhaps introduce an element of surprise.

In a published interview, Jonathon Porritt, the head of FoE England, readily acknowledged the pragmatic cooperation that occurs among environmental groups: "If any legislative initiative gets under way which will involve different groups, they will get together to determine tactics. This approach developed at the time of the Wildlife and Countryside Act, when all the groups worked out what they wanted to say and how they would approach the same target differently. Each used different tactics, but knew why the others were using their tactics" (cited in McCormick 1991, 37). The intra-movement alliance networks discussed in chapter 7 further attest to such cooperative activities. Both informal communication networks and formal participation in green umbrella groups enable environmentalists to maximize their political options.

One further advantage of this diversity in environmental action is that it allows ecology organizations to remain outside the structure of the prevailing sociopolitical order if they so wish and to represent the broader political goals of the movement. And yet, we find that the incentives to become engaged in conventional methods of political lobbying are so great that even ecology groups become active in these areas. In absolute terms, ecology groups are about as likely to engage in conventional lobbying strategies as they are to mount protests or other unconventional actions, though their conventional activities are often carried out in a quiet and almost surreptitious manner.[18] As they told us often during the interviews, protests alone are insufficient to change policy.

This pressure to get involved in conventional politics poses a dilemma for ecology groups and other New Social Movements. In his analysis of the peace movement, for example, Thomas Rochon (1988) concluded that there is a tension between the aspirations of NSMs for broad social change and the achievement of specific policy goals. The more a movement engages in conventional politics and the broader the attempts of formal and informal coalition building, the greater the likelihood of obscuring the fundamental social critique of that movement. In the case of the peace movement, the fundamental critique of militarism was overshadowed by the pursuit of a more immediate (and much more restricted) goal: the deployment of a particular kind of missile in particular locations among NATO members. Similarly, the anti–nuclear power movement has successfully attracted more support in opposing particular plants at particular sites (in the landmark cases of Whyl, in Germany, or Sellafield, in Great Britain) but less support for a global strategy of shutting down all nuclear power plants. As women's groups focus on activities that are necessary for the passage of legislation, the social-consciousness goals of the movement inevitably suffer (Ferree 1987).

To a large extent, environmental interest groups face the same dilemma, which is common to reformist movements: they must choose between being pragmatically successful (in broadening their popular base and reaching modest policy goals) and being true to their fundamental beliefs.[19] Indeed, we find that even ecology groups emphasize modest reformist goals in most of their activities. Perhaps the most prominent example of this institutionalization of the movement is the gradual moderation of the German Green Party, especially in its recent turn toward "realist" politics (Offe 1990; Frankland and Schoonmaker 1992). One of our more staunchly ideological respondents worried openly that "once you begin to participate in the game, then you have already lost."

It is too early to predict the outcome of this dilemma between pragmatism and ideology. In the final chapter, however, we shall examine how the resolution of this problem will be a key factor in determining the long-term significance of the environmental movement and its enduring impact on the politics of advanced industrial societies.

Chapter Nine

POLITICAL PARTIES

AND ENVIRONMENTALISM

One of the central research questions about the green movement concerns the partisan implications of environmentalism. The importance of this issue derives partially from the centrality of political parties in most Western democratic systems. Political parties provide the primary method of selecting political elites and largely determine the content of electoral and legislative agendas; in addition, the parliamentary structure of most European governments converts partisan majorities into control of the basic institutions of governance. Given this focus, discussions of interest-group politics must almost inevitably consider the relationship between interest groups and parties because of the predominate role of parties as institutions of articulation and aggregation.

An additional source of interest in the partisan implications of New Social Movements is the tumult and political change that most European party systems have experienced over the past decade (Dalton et al. 1984; Franklin et al. 1992). Several have become more fractionalized, and fluctuations in voting results increased at the aggregate and individual levels. Internal tensions and factionalism apparently are more prevalent, especially among parties of the Left. Many factors account for this new wave of partisan instability, but the contribution of environmentalism and other NSMs figures prominently in these partisan trends.

Several scholars maintain that the alternative political philosophy and political style of NSMs such as the environmental lobby are weakening the established political parties (Kitschelt 1989; Roth and Rucht 1987). The issues of noneconomic, collective goods espoused by the environmental movement cut across the dominant political cleavages of Western party systems—as illustrated by the visible controversies between New Left environmentalists and Old Left trade unionists in most of Europe's social democratic parties. In addition, the unconventional political style identified with NSMS is equally problematic to established party elites. The concepts of inner-party democracy, populist politics, rotation of office, direct-action techniques, and other aspects of the environmentalists' political repertoire seem incompatible with the oligarchic structure and corporatist tendencies that exist within most European political parties.

Examining the partisan orientations of the green movement therefore provides an important basis for determining the possible impact of environmentalism on European party systems. As the institutional base of the movement, environmental groups represent specific interests within the policy process and mobilize the public on green issues by providing political cues on which policies are important and which strategies—including partisan alliances—to adopt. In addition, for parties seeking to establish their own green credentials, environmental groups afford an institutionalized link between the movement and the parties. Although few formal ties now exist between political parties and environmental groups, the political perceptions of environmental groups and their informal party contacts indicate latent partisan tendencies that may eventually produce more enduring alignments.

In addition, the literature on New Social Movements increasingly emphasizes the role of political opportunity structures—in this case, potential partisan allies—in guiding the behavior of movement organizations (Kitschelt 1989; Tarrow 1989; Kriesi 1994; della Porta and Rucht 1994). In fact, because the pattern of opportunity structures is much clearer in dealing with partisan politics, one might argue that institutional structures will have a more direct impact on behavior than we found for other political actions in the previous chapter. By drawing upon the diversity of party systems in the ten nations studied, we can provide systematic evidence on the role of political opportunity structures on the partisan actions of environmental interest groups.

In this chapter we address the question of the partisan implications of environmentalism by going directly to the source—environmental interest groups themselves.

Three Models of Partisan Relations

The ability of a social movement to develop partisan ties presumes the availability of potential alliance partners. Whether such allies exist for the green movement, however, remains a matter of speculation. The recent history of Western European party systems makes it difficult, if not impossible, to identify a systematic pattern in the relationship between environmental groups and parties that transcends national boundaries. Many of the established political parties remain skeptical of environmentalists and the policy reforms they offer; other parties, of all political colors, are courting green voters with claims of a new environmental awareness. And everywhere, it seems, a plethora of New Left and green parties are enticing voters with their environmental programs. Rather than a guiding principle, the popular refrain among environmentalists that they "are neither Left nor Right" might be a reflection of this confusing partisan situation. Where, one might ask, are the real opportunities for partisan alliance that might structure political action?

This diversity of possible scenarios contains three basic models of how environmental groups might relate to the party system. Each model is somewhat visible in the relationship between environmental groups and political parties, though it may vary in frequency across nations and time; and each leads to somewhat different predictions about the long-term implications of the environmental movement.

One model involves alliances with an existing political party. Because of the structure of European party governance, there are many reasons to presume that existing parties would eventually incorporate any significant new political interest. Historically, European parties have integrated new social forces into the political system, leading to the formation of agrarian, religious, and labor parties. Indeed, Rudolf Heberle, an early scholar of social movements concludes that "in order to enter into political action, social movements must, in the modern state, either organize themselves as a political party or enter into a close relationship with political parties" (Heberle 1951, 150–51).[1] Given enough time, therefore, environmentalism and other NSMs might be integrated into the dominant partisan framework of most European political systems. Tarrow (1989) makes this argument for NSMs in Italy, and the refrain has been shared by others (Kriesi and van Praag 1987; Klandermans 1989).

In fact, the European environmental movement has had a tempestuous relationship with the major established parties on both the Left and the Right. During the mobilization phase of the environmental movement in the 1960s and early 1970s, many of the existing political parties quickly developed environmental programs. In the late 1970s and early 1980s, however, this partisan support for environmental reform waned, as a worldwide recession shifted the attention of most parties to other political agendas. Conflicts over policies to address these economic problems also highlighted inherent tensions between environmentalists and the traditional support groups of social democratic (labor-oriented) and conservative (business-oriented) parties. Environmentalists found themselves confronting an alliance of business interests and labor unions on issues such as nuclear power and the development of an industrial infrastructure (Siegmann 1985). Even if the established parties supported environmental issues while in opposition, they often proved to be unreliable allies once elected to office, at which point party leadership became primarily attuned to the policy demands of their dominant labor or business supporters.

Increased public support for environmental reform during the 1980s has elicited a renewed concern for the environment among established parties. Leftist parties have often taken the initiative in attempting to integrate environmentalists into their progressive coalition. The French Socialist Party, for example, has tried to develop new environmental credentials (Rohrschneider 1993c). Mitterrand courted environmentally oriented voters in the 1981 election campaign, though many environmentalists feel he reneged on his promises after winning the election with their support. When Premier Michel Roccard formed a new French government in 1988, he appointed the environmentalist Brice LaLonde to head the ministry of the environment.[2] Similarly, one of the most assertive groups in Italy, the Lega per l'Ambiente–Arci, was formed in the early 1980s by intellectuals from the Italian Communist Party (PCI) and until 1985 received financial support from the party's cultural organization. The Dutch Labor Party (PvdA) has been one of the most progressive parties in Europe on issues of environmental reform. In the 1987 national elections, the Society for the Defense of the Environment (VMD) provided voters with a scorecard on the environmental programs of the major parties, giving the PvdA the highest marks. Left-wing parties in Denmark, especially the small Socialist People's Party (SF) and Radical Venstre, have similarly competed in placing their green programs before the voters (Jamison et al. 1990).

The depth and persistence of leftist support for a green agenda remains uncertain, however. Hanspeter Kriesi (1994) points out that the French socialists drew votes from the green movement and then failed to deliver on campaign promises. The German Social Democrats (SPD) recently adopted a progressive environmental program, but only after being forced from government; while it was in the government, it was less supportive of environmental reform than its successors, the Christian Democrats (CDU/CSU). The commitment of Greek socialists (PASOK) to their environmental supporters lasted about as long as it took to count the votes for Papandreou's victory in 1981 (Stevis 1993).

Furthermore, environmentalism is not synonymous with the Left: parties across the entire Left/Right spectrum have attempted to co-opt or capture the environmental label as their own. In a theoretical discussion, Claus Offe (1985) outlined the potential alliance options of the environmental movement with both leftist and rightist parties. There is some evidence to support this argument. After the CDU's election victory in 1983, for example, Helmut Kohl moved aggressively to deal with acid rain, establishing new regulations for automobile emissions and pressing the European Community for European-wide regulations (Patterson 1989). Kohl also formed a new ministry of the environment and appointed a widely respected member of his party as head.

In Britain, each of the major political parties has an affiliated organization that focuses on environmental issues (the Conservative Ecology Group, Liberal Ecology Group, and the Socialist Environment and Resources Association [SERA]), even if these organizations lack significant status within their respective parties.[3] The Liberal Party and the Liberal/SDP Alliance were strong advocates of environmental protection, but institutional links to the environmental movement remained weak. Even Margaret Thatcher eventually claimed to be a born-again environmentalist in 1988—although most Britons were unconvinced. Thatcher's transformation from the iron lady to a green goddess is generally attributed to her awareness of the growing electoral potential of environmental issues. The Tory government soon introduced legislation ostensibly designed to strengthen pollution control standards and to restructure the administration of environmental protection policies, though these initiatives were generally criticized as insufficient by environmentalists. This dramatic birth of an environmental consciousness by Thatcher was unable to sway many British voters; the nascent British Green Party won 14 percent of the vote in the 1989 European parliamentary elections (Frankland 1990). John Major's accession as prime minister has probably improved the image of the Conservative Party on environmental issues, but perhaps only because it was so dismally low under Thatcher. Furthermore, although Thatcher at least attempted to court green voters, the British Labour Party seemed unreceptive to green thought. Temporary alliances have occasionally formed between environmental groups and the Labour Party, but these were short-lived and followed by feelings of incompatibility on both sides.[4]

Established parties in other nations also seem to have been paying more attention to environmental issues in recent years. In Holland, the liberal/ Christian government (VVD-CDA government) proposed a major new program of environmental reform in early 1989. Not to be outdone, the opposition parties, led by the PvdA, endorsed the goals of this legislation while adding counterproposals on funding the reforms that deadlocked parliament and brought about new elections. This was the first instance of a European government falling because of an environmental issue, and in the subsequent campaign environmental policies occupied center stage. Developments in Italy, including a national referendum on nuclear power and the emergence of a new green party, have caused the established parties to become more concerned about environmental matters (Diani 1990a).

In short, it is unclear how the green movement relates to the ideological spectrum of established European political parties. Although it might seem natural for a new political movement to seek alliances with an established participant in the party system, actual experiences make it unclear whether environmentalists see dependable allies among these parties. The commitment of established parties to environmental reform is secondary to their primary principles of representing other socioeconomic groups. Once in power, green promises often fade. In addition, the different styles of the green movement and party officials create a tension that causes further division.

Because of the difficulties in working with the established parties, many environmental activists (as well as those in other NSMs) claim their greatest opportunity for action is a second model of working with a new political party that advocates green and other New Left issues (Müller-Rommel 1989, 1993; Parkin 1989). This option assumes that environmentalists will be drawn into partisan politics because of the central role of party government in policy making throughout Europe. Rather than working through the existing parties, however, NSMs could create new political parties in their own image. Jonathon Porritt, former head of FoE, explains his activity in the British Ecology Party in these terms: "It is my contention that party political activity will always remain an essential part of that development [of green ideas in this country]; for better or worse, the Ecology party is the only organization prepared to take on that role to the fullest extent. But even as a political party, we have no illusions about the fact that our primary function is still an educative one, the spreading of green politics to as wide an electorate as possible" (Porritt 1984, 9). Green parties are not only new in chronological terms, they also represent parties of a new type (Kitschelt 1989; Kitschelt and Hellemans 1990; Müller-Rommel 1989). The German Greens provide a model of what we might expect: a highly politicized and issue-oriented party that adopts a decentralized, participatory structure in tune with the values of their supporters.[5] Green parties base their appeals on environmental issues and the goals of other NSMs, so they are not cross-pressured by the traditional social-group interests of the established parties. Researchers find that the greens and other New Left parties attract a distinct electorate, which increases electoral volatility and changes voting patterns by enticing more individuals to cross traditional party lines (Müller-Rommel 1993; Rüdig and Franklin 1992; Rohrschneider 1993b). The Greens also promote an unconventional style of political activity, as represented by their advocacy of protest and methods of direct democracy. As Petra Kelly of the German Greens once proudly proclaimed, the Greens are the "anti-party party."

Environmentalists and even environmental interest groups have frequently played an active role in the formation of new green parties. An association of several environmental groups, headed by AdlT, was instrumental in the creation of the French ecology party in 1978; for several more years AdlT remained involved in national and local elections in support of green candidates. The Belgian branch of AdlT was equally active in the creation of ECOLO, providing resources for the party and guiding its programmatic development. ECOLO first reached national prominence following the 1979 European parliament elections, in which thirteen of the seventeen ECOLO candidates were AdlT members. FoE activists were instrumental in creating the British Ecology Party (now the Green Party), and a strong personal network still links FoE and the present British Green Party (Porritt 1984). The Italian AdT also assisted in the formation of the Italian Green Party in the late 1980s. In other nations, environmental interest groups have avoided the organizational problems of creating a new party by establishing strong links to small New Left parties, such as the Radicals (PPR) in Holland or the Socialist People's Party in Denmark. By the end of the 1980s, green parties or their New Left supporters had seats in the national or European parliament in most of the nations we studied.[6] These legislative positions provide green-party delegates with a public platform for advocating environmental issues and challenging the programs of the government and the established parties.

Finally, others have argued that neither working with established parties nor creating a new green party represents a viable political strategy for the movement. A third apartisan model holds that environmentalists cannot (or should not) develop alliances with any political party and should instead work outside the established channels of partisan politics. Many environmentalists are hostile to political parties because of their different goals and different philosophy, that of maximizing their electoral appeal. Political parties pursue what Kitschelt (1989) terms the logic of electoral competition: attempting to aggregate political interests in the hope of increasing their vote share. Often these additional votes come from forsaking environmental issues to the interests of other constituencies. Environmental groups and other citizen interest groups are intent on articulating focused policy objectives, which may be diluted by such partisan aggregation. Environmentalists thus often view the parties as disinterested in (or opposed to) true environmental reform. Furthermore, the adherence of most political parties to the iron law of oligarchy produces a bureaucratic and hierarchic style of decision making that conflicts with the norms (and likely the political interests) of environmental groups. Consequently, Lowe and Goyder quote one British environmental leader as saying: "There's nothing to choose between them. Thank God the environment is not a matter of party politics" (1983, 72). This refrain, in some shape or form, was often repeated by the environmentalists in our sample. The placards of one French group were even more explicit in expressing disdain for parties and electoral politics: "Elections—piège à cons" (Elections—trap for idiots).

Past relationships between environmentalists and political parties often seem to validate this negative evaluation of partisan action; even the partisan relationships described in the preceding paragraphs were often brief and contentious. Ongoing, formalized ties between the movement and political parties are difficult to identify; often relationships are informal and based at a personal level. Environmental interest groups often insulate themselves from partisan politics, even when green parties are involved. FoE groups in Europe have been exceptional in their willingness to engage in electoral politics, having been instrumental in the formation of green parties in France, Belgium, Britain, and Italy. Most other green groups, however, have been visibly absent when new green parties were formed. For instance, official support from the major German environmental groups was surprisingly lacking at the creation of the Green Party in 1980; even activist environmental groups, such as the BBU, avoided formal endorsement of the party. Mario Diani (1990a) similarly describes how large Italian groups, such as WWF and Italia Nostra, consciously distanced themselves from the newly formed Italian Green Party. In short, an environmental interest group may attempt to maximize its political influence by presenting itself as a broad public interest group, without ties to a specific party.

These three models for relations between environmental groups and political parties differ greatly in the opportunities they imply for the environmental movement. Moreover, historical narratives on green action provide an uncertain basis for deciding among the alternatives outlined. Hard empirical evidence on the partisan affinities of the environmental movement is extremely rare. Analyses are often based on a subjective reading of party actions and electoral outcomes, or other information in the public record, without directly assessing the perceptual map of environmentalists. As we have noted, the lessons from this published record are uncertain as well as implicit. In spite of extensive speculation and insider accounts, we have little systematic evidence on actual ties between environmental interest groups and political parties in cross-national terms, nor do we know how these groups view partisan politics and the respective political parties in each nation. Our survey of environmental organizations provides some initial evidence to determine how environmentalists view the options available to the movement.

Partisan Friends and Foes

Before an interest group can or will develop ties to a political party, a common policy perspective must exist. Environmental groups must be able to identify their friends and foes and see a significant distinction between them.

To approach the issue of partisan allies in as neutral terms as possible, we asked group representatives the following: "Which political parties come closest in representing the interests of [your group]?" Our purpose was to identify political affinities, without asking for party endorsements.

Responses to this question on partisan perceptions were cumulated across the ten nations studied to determine whether a general pattern of partisan images exists (table 9.1). Reflecting the strong apartisan norms of the movement, the majority of groups initially react by stressing the apolitical or nonpartisan nature of their organization. Even with modest encouragement from the interviewer to name specific parties, and the assurance that we were asking not about formal ties but only about shared interests, almost one-third of our respondents could (or would) not identify a single partisan ally of their group. Environmentalists commonly explain these responses in terms of the explicitly nonpartisan nature of the group. In other instances, group officials state that because support for a specific party might alienate the voters of other parties among the membership, they avoid any expression of partisan preference.

Although the apartisan norms of environmental groups are strong, the data also illustrate that most environmental elites can identify sympathetic parties when pressed. Among those who named a specific party, the preference for leftist parties was overwhelming. At the time of our survey, viable green parties existed in only three nations (Germany, Belgium, and France).[7] Within these three nations, nearly all the environmental groups saw the greens as supporting their cause. Various small New Left parties exist in almost

TABLE 9.1

Parties Representing Interests of Environmental Groups

Party Type	Conservation Group	Mixed Group	Ecology Group	Total
Green	19%	25%	24%	21%
New Left	24	13	71	38
Communist	5	13	5	8
Socialist/Labor	24	25	38	29
Liberal	22	25	38	30
Conservative	3	0	0	2
Other	0	0	5	2
All parties	8	13	5	8
No party/Don't know	32	38	0	23
Refused to respond	5	0	10	6
Total Responses	142%	152%	196%	167%
(*N*)	(37)	(8)	(21)	(66)

Note: Figures are the distribution of responses to the question: "Which political parties come closest in representing the interests of the group?" Because multiple responses were possible, columns may total more than 100%.

every nation, making these parties the most frequently mentioned partisan ally of the movement across Europe (38 percent). Several environmentalists also cite socialist/labor parties and liberal parties as potential supporters, although evaluations of these parties vary widely across national boundaries. Conservative parties are the obvious void in the potential alliance network. Of all the groups we surveyed, spanning ten nations and a range of environmental orientations, there was only one mention of a major conservative party. Although several conservative party officials have attempted to redress the negative image of their parties, environmentalists fail to see conservatives as supporting their interests.

The other dimension of partisan images involves opponents.[8] Again reflecting a tendency to avoid expressions of partisan sentiments, nearly half of the groups decline to mention a specific party as disinterested or hostile to the group's cause (table 9.2). When environmentalists name a potential partisan opponent, they seldom cite leftist parties and instead target conservative parties. In broad cross-national terms, therefore, environmental interest

TABLE 9.2

Parties Opposing Interests of Environmental Groups

Party Type	Conservation Group	Mixed Group	Ecology Group	Total
Green	3%	0%	0%	2%
New Left	0	0	0	0
Communist	5	13	10	8
Socialist/Labor	0	13	5	5
Liberal	8	25	33	18
Conservative	19	63	57	36
Other	3	0	5	3
All parties	8	8	0	6
No party/don't know	57	13	29	42
Refused to respond	3	0	14	6
Total Responses	106%	135%	153%	126%
(N)	(37)	(8)	(21)	(66)

Note: Figures are the distribution of responses to the question: "Are there parties that you feel are disinterested or even hostile to your cause?" Because multiple responses were possible, columns may total more than 100%.

groups do seem to be loosely integrated into the traditional Left/Right framework of the European party systems.[9]

As with other elements of environmental action, we expect partisan activities to be influenced by the environmental ideology of a group. In their role as challenging groups, we might expect ecology organizations to exhibit great disdain for partisan politics as now constituted. Partisan politics inevitably involves compromise and the adjustment of one's goals to a party's other interests. Moreover, NSM researchers have stressed the antipartisan tendencies of these movements. Yet partisan politics is also a route to political influence, and ecology groups have been more active than conservation groups in forming green parties and in providing electoral support for established parties.

In terms of their basic orientation toward partisan politics, we find that ecology groups are less likely than conservation groups to take an apartisan stance (table 9.1). Most ecology groups (85 percent) identify at least one specific party as representing their interests, and most also name a potential partisan opponent (57 percent). In contrast, barely half (55 percent) of the conservation groups mention a specific party that represents their interests, and less than half note a partisan opponent. For the most part, ecology groups find their partisan representatives outside the circle of established political parties, primarily in small green and New Left parties. But because there were only three green parties at the time—the Greens (Germany), ECOLO/ Agalev (Belgium), and Ecologists (France)—they are mentioned infrequently. Less than half of the ecologists mention either leftist or liberal parties as reflecting their views, and none mention a conservative party.[10]

Environmental Performance

As we saw in chapter 7, the environmental performance ratings of political parties yield a similar pattern across party groups. We presented environmental leaders with a list of social and political organizations, including several political parties from their nation, and asked them to evaluate how good a job each organization was doing in addressing the environmental problems in that nation (see fig. 7.1). Green parties and small green-oriented parties receive the highest overall environmental scores averaged across all nations (mean score = 3.05). Established parties generally receive negative ratings, and only modest differences emerge between the major party groupings. The "family" of European socialist parties receive a 2.02 rating on average, compared to 1.66 for liberal parties and 1.55 for conservative parties. The differences across these three established party groups thus barely spans a half-point on a four-point scale, far less than the difference of one point or more separating green and New Left parties from the three groups of established parties.

The broad cross-national pattern inevitably becomes more complex when one examines the images of specific parties within each nation (fig. 9.1). We caution the reader that these comparisons are based on a small number of groups per nation; and we interviewed only one spokesperson from each group (Ireland and Luxembourg are not presented because we contacted only two groups in each country). Still, relative party rankings are fairly consistent across groups with different ideologies and policy interests. Furthermore, these groups are not just a sample but make up the relevant population of national environmental interest groups for that country.

The data in figure 9.1 highlight environmentalists' broad criticism of the major established European parties. Green and New Left parties, such as the Danish Socialist People's Party (SF), the Italian Radicals, or the Greek Eurocommunist Party (KKE-Interior), systematically receive positive policy

FIGURE 9.1

Perceptions of Environmental Performance of Political Parties, by Nation
Environmentalists rated the major political parties in their country using the
following scale: 4 (excellent), 3 (good), 2 (fair), and 1 (poor).

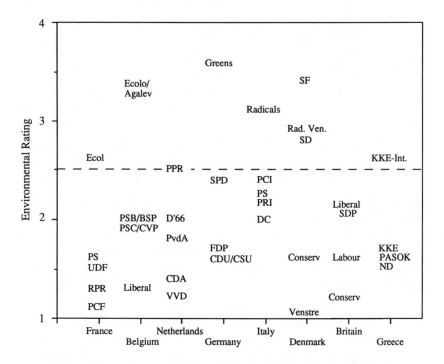

endorsements from environmentalists. Only the Danish Social Democrats
(SD) are rated above the midpoint on the performance scale, while twenty-
seven other large established parties rank below the midpoint.

The data also illustrate the extent to which environmental policy is seen
as cutting across the traditional Left/Right dimension of European party
systems. The perceived differences across the established parties are normally
modest, whereas the gap between the green or New Left party and the nearest
established party is generally larger. Thus, the major environmental cleavage
runs between the green/New Left parties and all other parties.

Even the rankings of established parties seldom conform exactly to a
traditional Left/Right party ordering. French environmental elites, for example, gave the lowest rating to the Communist Party (PCF)—even below

those for the conservative parties (RPR and UDF)—because the communists' pro-growth philosophy has led the party to oppose strict environmental protection as a threat to the economic interests of the working class. Environmentalists were critical of most liberal parties: the liberal parties in Belgium (PLP/PVV), the Netherlands (VVD), Denmark (Venstre), and Germany (FDP) receive among the lowest environmental ratings in their respective nations. Only the British Liberals receive positive evaluations relative to the other parties. The business and agricultural orientations of liberal parties often lead them to become prime critics of the environmental movement, despite their liberalism on other social issues. These same conservative economic tendencies exist within the major conservative parties in Europe, but they are partially moderated by the broad catch-all nature of these large parties. The British Conservative Party, for instance, must be partially responsive to the demands of conservationists and protectors of the English countryside, who make up a significant share of the party's traditional constituency. Similarly, the policies of the German Christian Democrats are most visible on environmental issues that affect their rural or middle-class supporters.

The potential for building alliances with established parties is probably greatest with the social democratic parties of Europe. Leftist parties are the normal electoral outlet for new progressive movements, and most European socialist parties are developing an interest in the issues raised by the environmental movement—for the simple reason that green and New Left parties are making inroads into the socialists' electoral support among young, university-educated, postmaterial voters. Still, environmental elites generally do not give social democratic parties high marks for their policy performance. In France, Belgium, and Greece, environmentalists give the Socialist Party a rating that is virtually indistinguishable from the rival conservative party. Only in Denmark and Germany do the socialist parties receive markedly better environmental ratings than the other established parties. Across the map of Europe, green interest groups do not perceive the social democratic Left as their obvious and strong ally.

In summary, the leaders of the groups we interviewed do not see that the issue of environmentalism is producing clear and consistent allies among the established political parties. In this case, choosing an ally is like choosing the best from among the worst. Moreover, party differences on the environmental dimension do not necessarily follow a traditional Left/Right division. The ability of environmentalism to produce a major realignment across European party systems therefore appears somewhat attenuated, at least until clearer lines of partisan differentiation develop. At the same time, it may be premature to accept the self-proclaimed apartisan label of environmental groups. Representatives of ecology groups, for instance, initially declared their apartisan tendencies during our interviews, but these assertions were followed with specific party evaluations. In fact, most environmental elites can identify some potential ally among the parties competing in their nation— usually a green or New Left party. From the perspective of the green lobby, the primary partisan outlet of environmentalism occurs outside the structure of major established parties. The ability of socialist or liberal parties to generate positive responses is clearly secondary to environmentalists' support of alternative parties. In spite of the theoretical alliance between conservative parties and green groups that Claus Offe (1985) postulates, in real terms these tendencies are nonexistent.

Working with the Parties

In order for the environmental movement to have an impact on party alignments, environmental groups must actually work with and through parties to represent their interests, establishing the working relationships that provide clear political cues for the public and lead to the alliances that typify the relationship between parties and labor unions, business associations, or agricultural lobbies. Alternatively, the apartisan rhetoric of environmentalists may preclude direct party contact, at least for the present.

In spite of the supposedly antiestablishment image of the environmental movement, we found that most environmental groups are in direct contact with government policy makers (see table 8.1). More than 50 percent of the groups surveyed maintain frequent formal or informal contacts with government officials and parliamentary committees dealing with environmental issues. In addition, more than 40 percent of the groups participate often in government commissions and advisory committees—the epitome of corporatist politics.

Against this backdrop of considerable interaction between environmental interest groups and the institutions of government, the low level of partisan activity stands out sharply. Environmentalists cite contact with political parties less frequently than any other activity on the list. Barely one out of ten groups claims to have frequent contact with party leaders (see table 8.2). Similarly, an open-ended question about each group's specific political activities over the preceding year displays an equal paucity of party contacts; only 6 percent of the groups mention party contacts as part of the group's major policy initiative (table 8.1). These findings present what we feel are the strongest evidence of the apartisan nature of the environmental lobby.

Although environmental groups are able to identify potential partisan allies and foes, they nevertheless shun direct party contact.

Explaining Partisan (In)action

How can we account for environmentalists' aversion to partisan activity? One of the most convincing explanations lies in their general skepticism about the effectiveness of party work. We heard repeated tales of the unreliability of party promises and the futility of past partisan activities. Virtually all the groups that had formally been involved in electoral politics recounted negative experiences similar to those of the British FoE noted earlier in this chapter. Problems exist across the political spectrum: established parties too often prove to be unreliable political allies, and the small green and New Left parties are difficult to work with and have limited electoral appeal. Even as a general strategy, environmentalists doubt the efficacy and reliability of partisan activity. French and Greek environmentalists were alienated by having supported the socialists, only to have Mitterrand and Papandreou implement different policies when elected. Mitterrand had voiced skepticism of France's aggressive nuclear power program during the 1981 election, but after a brief debate the new socialist government chose to support a program of nuclear power expansion that included orders for six new reactors in 1982–83, completion of the Super-Phoenix breeder reactor, an expansion of nuclear processing capacity, and a long-term goal of increasing France's reliance on nuclear power (Hatch 1986, chap. 6). Many Greek environmentalists had close ties to PASOK and expected the party to advocate green reforms after it won control of the government in 1981. But Papandreou's government ended up stressing economic goals over environmental protection, and none of the other established parties stepped forward to claim the green agenda (Stevis 1993). Dutch groups expressed their political misgivings with similar stories of PvdA and D'66. Environmental groups have learned that they can influence parties and trust their pronouncements only until the ballots are counted.

A specific question in the survey substantiates these impressions. We asked group representatives about the effectiveness of various methods as a means of influencing government policy. A majority of the groups (55 percent) declare that working with parties is rarely or never an effective method of policy influence (see fig. 8.1).[11] Furthermore, the perceived effectiveness of party activity is below other policy activities, such as working with the media, the environmental ministry, or governmental commissions.

Relating party activity to perceptions of the effectiveness of this activity is, however, a somewhat circular argument. The apartisan tendencies of

TABLE 9.3

Correlates of Party Contact

Organizational characteristics

Age of group	−.25*
Size of professional staff	−.09
Adequate staff size	.03
Size of membership	.18*
Annual budget	.04
Mass-membership group	.00
Leadership control	.03
Membership input	−.02

Opportunity structures

Viable green party	−.09
Viable New Left party	−.16
Leftist party in government	−.22*
Corporatism	−.09
Openness of system	.07
Evaluations of parliament	.08

Organizational values

Ecological orientation	.13
Left/Right ideology	−.06
Satisfaction with democracy	−.04

Note: Table entries are Pearson correlation coefficients. Coefficients marked by an asterisk are statistically significant at the .10 level.

environmentalists may preclude objective assessments of alternative political techniques. We would expect that calculations about partisan politics, much like those for conventional and unconventional policy action, are conditioned by the characteristics of interest groups and by the external political environment.

Using some of the measures first presented in chapter 8, we analyzed the role of organizational characteristics, opportunity structures, and organizational values in affecting the partisan behavior of green groups (table 9.3).[12] As we determined in chapter 8, large and well-financed groups possess the resources to participate in a wide range of political activities; this holds true especially for conventional activities and to a lesser extent for unconventional tactics. In this case, however, we find that organizational characteristics have relatively little impact on the level of partisan activity. Groups with a large membership base are slightly more likely to contact political parties, but this relationship does not hold for other resource characteristics (such as staff size or annual budget). In fact, the only statistically significant correlation is with the age of a group; recently formed groups are more likely to participate in partisan politics.

Resource mobilization research suggests a second class of variables—institutional context—as an important factor in the political activities of SMOS. Herbert Kitschelt applied his earlier (1986) analysis of opportunity structures to his work on green parties in Germany and Belgium (Kitschelt 1989; Kitschelt and Hellemans 1990). Hanspeter Kriesi (1994) builds upon this model, implying that opportunity structures should affect partisan activity since the choice of political allies and the focus of partisan action is relatively clear-cut.

We assembled data to test this model. One might assume that the existence of a potential partisan ally is a first prerequisite for partisan action. We therefore identified nations with a viable green and/or New Left party, expecting to find higher partisan activity there. Analysts have also discussed the role of the major leftist party in stimulating NSM activity. Kriesi (1993) suggests that a leftist government will decrease the general political mobilization of NSMs, because movement organizations will establish political alliances with the leftist government and win policy reforms. Others, such as Donatella della Porta and Dieter Rucht (1994), suggest the opposite; that major leftist parties are more responsive to NSMs when the party is outside of government and thus does not have to deliver on their promises for policy reform. We can test the validity of these rival claims. Finally, there has been a general presumption in the RM literature that alliances between NSMs and established political channels are more likely to develop in open political systems rather than in systems with closed corporatist structures (Kitschelt 1986; Kriesi and van Praag 1987).[13] Although not directly linked to partisan activity, this assumption can easily be extended to contact with the established political parties: in open systems, one or more of the established parties is likely to develop ties to new contenders, such as environmentalists or women's groups.

The second panel of table 9.3 lists the correlates of national political structure and partisan activity. Surprisingly, the availability of viable partisan allies does not facilitate party contact by environmental groups. In fact, the relationships are opposite what RM theory would suggest (though they are statistically insignificant). Frequent party contact is slightly lower (12 percent) in the three nations with a viable green party at the time of our survey than in those without one. There is also a weak negative relationship between the presence of a New Left party in a nation and partisan activity by environmental groups. Leftist control of the national government similarly seems to discourage partisan activity by environmental groups. This finding might be explained by the specific nations that had leftist governments in 1985–86 (France, Greece, and Ireland), though it mirrors a common feeling that leftist parties are more willing to work with environmental groups when they occupy the opposition benches.

In addition, the broader opportunity structures of a nation—represented by the extent of corporatist policy making and an open policy style—have a negligible impact on partisan activity. Even perceptions of the environmental performance of parliament, a surrogate for psychological perceptions of the openness of the system, have little impact on partisan activity. All of the relationships involving opportunity structures are weak and consistently suggest that even when environmental groups have access to sympathetic political parties, they remain outside the sphere of partisan politics. Because partisan activity is more concrete than the broad measures of conventional and unconventional action examined in chapter 8, and because the opportunity structures for partisan action are more distinct, the lack of effect provides yet further evidence that opportunity structures are of little consequence in guiding the behavior of interest groups.

It is unclear how our framework of Ideologically Structured Action should apply to partisan politics. In the previous chapter we showed that ecology groups are substantially less likely to participate in conventional forms of policy influence and more likely to utilize unconventional tactics. Based on these indications, one might expect ecologists to shun party contacts. Alternatively, we have cited anecdotal evidence that ecology groups are more likely to participate in electoral politics and have played a significant role in the formation of several green parties.

Our data show that ecology groups are more likely than conservation groups to work with political parties, although this difference is not statistically significant. Only 6 percent of conservation groups contact party leaders often, compared to 20 percent of ecology groups ($r = .13$).[14] Many ecologists are sharply opposed to partisan activity, among them several Greenpeace organizations and some of the more decentralized and unconventional ecology groups. At the same time, some of the most visible examples of formal working relationships between environmental groups and political parties have involved ecology groups. In most instances, these organizations have focused their efforts on supporting green or New Left parties, as in the case of FoE in several European countries. Other organizational values, such as Left/Right ideology or satisfaction with democracy, are virtually unrelated to partisan activity.

We entered the variables from table 9.3 into a stepwise multiple regression model to determine which factors have the greatest independent influence in predicting partisan activity.[15] The age of the organization is the first variable to enter the model ($\beta = -.37$), followed by membership size ($\beta = .32$). Neither opportunity structures nor organizational values entered into this model. Overall, the two predictors have a relatively weak impact on participation patterns (Multiple $R = .39$).

The lesson to be learned from these analyses is that a group's willingness to participate in partisan politics is relatively unaffected by the external environment or the characteristics of the group. Opposite to the pattern for media contact, avoidance of partisan activity seems widespread among environmental interest groups. Environmentalists say that political parties operate under a Weltanschauung that prevents close ties between group and party. The spokesperson for one Greenpeace organization, for example, stated that "the philosophy of the parties is totally different, they have only a short-term, pragmatic perspective." Sentiments of this sort were interspersed throughout our interviews. Environmentalists feel that parties prefer to deal with green issues in an incremental way and want to avoid the fundamental questions raised by environmentalists—especially ecology groups. Thus parties become part of the problem rather than the solution. Other respondents complained that electoral politics and the logic of party competition encourage parties to water down their positions and concentrate their efforts on mobilizing potential voters. A member of a German group maintained that the political parties do not want to support the environmental movement; instead they attempt to coerce environmentalists into supporting the party. On the whole, group representatives believe that they exercise a political role that is partially in conflict with the norms of partisan politics. Furthermore, because these views are common among environmentalists, the structure of short-term political opportunities has very little impact on the potential for partisan action.

Patterns of Voting Support

A major justification for examining relations between environmental groups and political parties involves the potential electoral consequences of environmentalism. Contemporary electoral systems are in a state of flux, and the new issues of advanced industrial societies, such as environmental protection, are both contributors to this partisan instability and a potential source of partisan alignment (Inglehart 1990; Franklin et al. 1992).

The potential impact of the environmental issue on voting is considerable. Green parties are displaying remarkable electoral strength throughout Europe (Müller-Rommel 1989, 1993). Moreover, many more Europeans express a willingness to vote for an environmental party—the potential electorate for a green party rivals that of socialist and Christian democratic parties (Inglehart 1990, 266)! Many American, British, and German voters openly state that a party's environmental policy would influence their voting decisions (Milbrath 1984). As green parties have made inroads into some of the traditional bases of electoral support for established parties, these parties have been forced to respond with their own environmental programs. The prominence of environmental issues was most apparent in the 1989 European parliament elections, when parties of all colors—from the Italian Communists to German right-wing Republican Party—included environmental statements in their party programs.

It is difficult to predict how Western parties will resolve the electoral opportunities and challenges represented by the environmental movement, but we can illustrate the present relationship with public opinion data taken from a Eurobarometer survey paralleling our group study. The Eurobarometer of April 1986 asked Europeans about their attitudes toward ecology and nature conservation groups (see chap. 3 for additional analyses of these data). The questionnaire did not tap support for specific environmental groups, so the reference structure used by respondents is likely to vary across individuals, and especially across nations. The term *ecology movement* undoubtedly evokes many different images. We therefore expect that this question attenuates the actual impact of environmental attitudes on party preferences (for more evidence on this point see Rohrschneider 1993b and Cotgrove 1982). A brief analysis of these data can nonetheless illustrate the general relationship between support for environmental groups and the voting pattern of European electorates.

We determined the percentage of voters of each party drawn from the ecology movement, whether actual or potential members of a group (fig. 9.2).[16] Because of the large differences in the size of ecology groups cross-nationally, the statistics in the figure are presented in terms of deviations from the national average.[17]

Green and New Left parties are clearly the most dependent on ecologist voters. For instance, 83.1 percent of the voters for the West German Greens

FIGURE 9.2

Ecologist Voters as a Percentage of Party Constituency   *Source:* Eurobarometer 25
(April 1986)

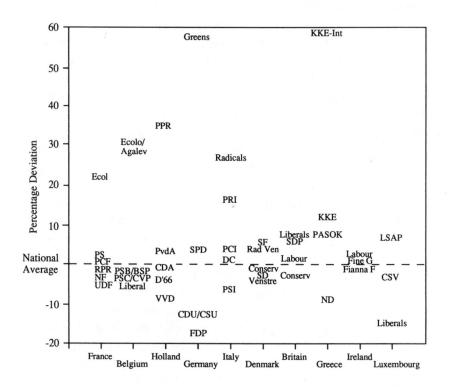

are members or potential members of an ecology group, which is 57.4 percentage points higher than the share of ecologist voters among the entire electorate. The French Ecologist Party, Belgian ECOLO/Agalev, Dutch PPR, Italian Radicals, and Greek KKE-Interior are also composed of a disproportionate number of ecologist voters.[18]

Ecologist voters comprise a much smaller percentage of the electorate for the established parties, though patterns of voter support seem to mirror traditional Left/Right alignments more clearly than environmentalists' evaluations of the parties (compare fig. 9.1). In every country but Denmark, the major leftist party has a larger ecologist constituency than the major rightist party—though these differences are often quite modest. The lack of ecologist support for liberal parties further blurs the traditional Left/Right party alignment. In Belgium, Germany, Denmark, and Luxembourg, the liberal parties

TABLE 9.4

Correlation between Environmental Group Support and Party Preference, 1982–89

Nation	Ecology Groups				Nature Protection Groups			
	1982	1984	1986	1989	1982	1984	1986	1989
France	.20	.19	.19	.18	.13	.14	.15	.17
Belgium	.20	—	.20	—	.17	—	.18	—
Holland	.18	.17	.17	.15	.15	.12	.16	.14
Germany	.23	.22	.24	.21	.13	.13	.10	.15
Italy	.10	.12	.12	.14	.11	.17	.11	.15
Luxembourg	.15	—	.13	—	.16	—	.14	—
Denmark	.21	—	.20	—	.14	—	.14	—
Ireland	.09	—	.12	—	.11	—	.12	—
Britain	.11	.12	.13	.13	.10	.09	.10	.13
Greece	.13	—	.29	—	.15	—	.19	—
10-nation Average	.16	—	.17	—	.14	—	.14	—
5-nation Average	.16	.16	.17	.16	.12	.13	.12	.15

Source: Eurobarometer 17 (April 1982), 21 (April 1984), 25 (April 1986), and 31A (May 1989).
Note: Figures are the Cramer's V correlations of party preferences with support for ecology groups and nature protection groups.

attract the smallest proportion of ecologist voters. The major environmental cleavage based on constituency, however, is between green or New Left parties and all of the established parties.[19]

In overall terms, the patterns of party support from ecologist voters (fig. 9.2) closely mirror the parties' environmental rating by group representatives (fig. 9.1). Party scores between the two figures are correlated at .69! Although this is not evidence of a causal relationship, it does signify the convergence in party perceptions by environmental elites and European publics.

This emerging picture can be expanded. Using four Eurobarometer surveys to track voting patterns across the 1980s, we examined the correlation between voting preferences and support for ecology and nature protection groups.[20] Because of the ambiguous ordering of parties in terms of their environmental image, we calculated the simple nominal relationship (Cramer's V correlation) between environmental group support and party preference (table 9.4).

Our data suggest that ecology groups are a more partisan symbol of the environmental movement. Adherence to an ecology group is strongly related to voting patterns in several nations—Germany, Belgium, Denmark, and France—probably because ecology groups are more involved in partisan politics in these nations and are more electorally relevant to voters. Attachments to a nature protection group yield similar patterns of party support, but the differences in voting behavior are more muted, because conservation groups are generally less relevant to partisan politics.

Environmental interests and party preferences converge most strongly in Germany and Denmark. As most German politicians readily acknowledge, the growing electoral impact of the Greens has forced all the parties to focus on environmental issues, and these issues are playing an increasing role in voters' electoral decisions (Kuechler 1991; Rohrschneider 1993b). In Denmark, the environmental banner is carried not by a green party but by several small leftist parties. Consequently, the environmental cleavage is already integrated into the traditional Left/Right framework of Danish politics, and support for the environmental movement is strongly related to vote choice (Anderssen 1988; Pedersen 1989).[21] Attachments to ecology groups are moderately related to party choice in France, Belgium, and the Netherlands, countries in which small green or New Left parties have advocated environmental issues and forced the established parties to respond at least partially to these new concerns. In these cases, therefore, the partisan integration of the environmental cleavage into voting patterns is linked to a new party emerging to represent this issue within the electoral arena.

The partisan impact of the environmental cleavage is smallest in Mediterranean Europe and the British Isles, where the correlation between support for ecology groups and party preferences generally stays below .20. The weak connection in Greece, Italy, and Ireland probably represents the limited salience of the issue to these publics (see chap. 3). All three nations have a low level of economic development and weak environmental movement. Britain, though, does not fit this pattern: it is one of the most highly industrialized and urbanized nations in Europe, with an exceptionally large and active network of environmental groups. We would trace the limited electoral impact of environmental attitudes in Britain to the established parties' hesitancy to adopt credible environmental programs and to the ability of the British electoral system to exclude new parties from representation (Rüdig and Lowe 1986). Even the dramatic breakthrough of the British Greens in the 1989 Europarliament elections appears to be a momentary success (Rüdig and Franklin 1992). Environmental voters in Britain simply lack a clear and dependable partisan ally.

Our overall findings suggest that the environmental cleavage is being integrated into the framework of several European electoral systems, although there is significant cross-national variation in this relationship. It is difficult, however, to judge the causal direction of this relationship. Because environmental interest groups generally avoid explicit partisan activity, this relationship must be due at least as much to the ability of parties to attract environmentally interested voters as to the ability of environmentalists to guide voters to a preferred party. Moreover, the process of integration is preceding very slowly. In spite of the very visible gains by green and New Left parties during the 1980s, the overall impact of environmental orientations on party preferences barely changed during the decade (table 9.4). Environmentalism remains a potential basis of division rather than a new framework of party alignment.

The Implications for Western Party Systems

We began this inquiry by presenting three alternative models to describe the relationship between the environmental lobby and political parties in Western Europe. One model suggested that environmentalists could become integrated within the established parties, primarily among leftist and centrist parties. A second postulated that environmentalism could provide the basis for parties of a new type—green and New Left—that could carry the values and unconventional style of the movement into the party domain. A third proposed that environmental groups would adopt an apartisan strategy, remaining aloof from partisan politics. In different ways, our findings provide some evidence in support for each model.

Instead of uncovering clear partisan tendencies, our research underscores the diversity of partisan orientations among environmental groups. Various groups see potential allies among green/New Left parties, Old Left parties, or liberal parties. Furthermore, there are different patterns of partisan affinities in virtually every European party system. In one nation, environmentalists may find the liberal party the most sympathetic to their views; in another it may be the socialist party or a New Left party. Moreover, the perceived differences in the environmental sympathies of the established parties are often fairly minor and not sharp enough to produce major electoral change. In spite of the new-found attention to environmental issues by established parties, we doubt that the patterns we described have changed significantly since our survey (e.g., see Rohrschneider 1993b; Rüdig and Franklin 1992). The lack of a consistent fit between the environmental issue and existing partisan alignments thus mitigates the possibility of a wide-ranging, systematic partisan realignment.

The uncertain partisan effects of environmental interest groups are further attenuated by the pattern of interaction between the movement and the parties. Although most environmental groups can identify friends and foes, they nevertheless adhere to their apartisan rhetoric in avoiding formal association or even direct contact with party elites. Although environmentalists inevitably deal with party leaders in their official positions as legislators or government administrators, they consider direct contact with the parties contradictory to the goals of the movement—to a degree, this includes even formal contact with green or New Left parties. This apartisan sentiment appears to be so deeply ingrained that, according to one of the central figures of the British environmental lobby, all interest groups in Britain must avoid party ties to be effective in influencing policy. When we brought the cases of the Trade Union Congress or Confederation of British Industry to his attention, he appeared to view these examples as novel ideas.

The greatest potential for significant electoral change by environmental groups, if it exists at all, seems to be in relation to green and New Left parties. First, surveys generally indicate widespread popular support for strict environmental policies and, in principle, strong support for environmental parties (see chap. 3). Second, the major line of partisan cleavage on the environmental issue runs between green/New Left parties and all other parties. Yet the potential electoral appeal of green parties is lessened by their unconventional style and often extremist views. In addition, the apartisan tendencies of environmental groups weaken the electoral potential of even the moderate green parties. Environmental groups are almost uniformly hesitant to establish firm and direct ties to any political party. Unless environmental groups break from their aversion to party contact, their partisan impact is likely to remain fluid and uncertain.

Even if political alliances were possible, environmentalists are skeptical about how they would enhance their political influence. Group leaders maintain that too close an identification with any one party would undermine their influence as a public interest group. They worry that the parties would co-opt them and that they would lose their clout as a single-issue lobby; experience tends to confirm these views. In addition, environmentalists feel that partisan allies would inevitably create partisan enemies. The gain of influence when one's preferred party was in power would not compensate for the lack of influence when the group's party sat on the opposition benches.

Environmentalists believe their strength lies in the popular base of the movement and the ability to pressure governments of all colors; partisan politics would only narrow public support for environmental groups and restrict their access to decision makers. The head of one British environmental group illustrated this point by explaining how they would pursue issues that were attractive to Tory voters when the Conservatives were in power (for example, the Channel Tunnel and countryside protection) and another agenda if the Labour Party assumed office (exposure to toxins in the workplace and urban decay). They can mobilize public support for these campaigns only by appearing to remain above partisan politics and by representing the public interest before both parties. To stay above partisan politics is thus an advantage for broad-based public interest groups.

If environmental groups fail to establish clear partisan ties, the political parties may still act to integrate the environmental cleavage into partisan politics. The head of one British group, who stood for parliament in 1987, acknowledged the hesitancy of environmentalists to engage in electoral politics and said that his party (the SDP) would therefore have "to capture the flag of environmentalism" on its own. Several political parties in Europe have been actively working to develop green credentials as voter interest in these issues has increased. These efforts, however, often display the internal party tensions that environmental issues evoke within the established parties, leading the parties to advocate an ambivalent composite of environmental policies. Contemporary electorates were understandably leery when Margaret Thatcher and George Bush declared themselves environmentalists. Even party elites who are openly sympathetic to the environmental movement may find themselves constrained by their party's traditional constituencies.

The uncertain implications of environmentalism for Western party systems are aptly illustrated by our comparisons of ecology and conservation groups. As vanguards of the environmental movement, ecology groups should accentuate the political tendencies of the movement. We have noted, however, that our model of Ideologically Structured Action has ambiguous predictions for partisan activity. There are plausible reasons for expecting that ecology groups might follow any one of the three paths outlined. In fact, we find that ecology groups illustrate the conflicting patterns of action we have described in this chapter. Ecology groups initially are the most vocal in proclaiming their apartisan orientations but are also the most interested in influencing policy outcomes. In the European political context, policy influence must involve some degree of party contact. Thus, ecology groups are slightly more likely to engage in party-oriented activities, while simultaneously criticizing the parties and the system of party government. Ecologists have not entirely yielded to the system of partisan politics, however; party contacts remain informal rather than institutionalized, and private rather than public. In short, the rhetoric of ecologists further obfuscates the activities that are occurring behind the scenes.

For the present, environmentalism and green issues are contributing to the general weakening of party alignments in Western democracies. Environmentalists are forcing green issues onto the political agenda of most industrial democracies, but they avoid linking this issue to the ongoing political alignments of Western party systems. That is, they are telling voters to act on green issues—but then failing to tell them how to act in partisan terms. This pattern may be a unique feature of the environmental movement, reflecting the broad popular base of the movement and the noneconomic, collective nature of the environmental issue. Yet a similarly ambiguous pattern of partisan interaction has been found in studies of the European peace movement (Rochon 1988) and the women's movement (Gelb 1989; Ferree 1987).

Even if political discourse is able to move beyond traditional economic issues to the issues of collective goods advocated by public interest groups, modern party systems may have difficulty representing and articulating these new issues. Our findings may, therefore, illustrate a pattern of fluidity and diversity that will increasingly characterize the style of citizen politics in advanced industrial societies. If citizen action groups and public interest groups continue to function outside the partisan arena, both the role of parties as articulators of political interests and the overall system of party government in Europe will be weakened. Political parties may eventually have to share the political stage with these new actors, who will challenge party dominance in policy making. The long-term impact of the environmental issue may be in contributing to a more fluid system of interest representation in advanced industrial democracies rather than in creating new partisan alignments.

5

CONCLUSION

Chapter Ten

ENVIRONMENTAL POLITICS AND

ADVANCED INDUSTRIAL DEMOCRACIES

Writing a decade ago, the American sociologist Robert Nisbit observed, "It is entirely possible that when the history of the twentieth century is finally written, the single most important social movement of the period will be judged to be environmentalism" (1982, 101). That observation is even more appropriate today.

In the 1980s environmental issues emerged onto the center of the political stage in virtually all advanced industrial democracies, and they remain there in the 1990s. In this study we have documented the extensive popular base for environmental reform in most Western European countries (chap. 3), findings that easily apply to North America as well (Dunlap 1992; Gallup Institute 1992). One can speak of a fundamental shift in citizen values when a majority of Europeans and Americans prefer strict environmental regulations over policies aimed at ensuring continued economic growth. Although environmental issues were once stigmatized as being of interest only to radicals or eccentrics, they now play a significant role in the policy decisions of governments and the life-style choices of many individuals. We are experiencing a greening of Western society.

Much remains to be done, but during the past decade there has been a general expansion of legislation dealing with environmental problems and, more important, an integration of environmental issues into the regular policy process of most European democracies (Vogel 1988; Kamienicki 1993; Commission of the European Communities 1990; Judge 1993). At the national level, even once-hostile governments are taking a more activist stance toward the environment. In addition, an expanding network of international agreements and decision-making structures is hastening the spread of environmental protection (Caldwell 1990, chap. 8; Haas et al. 1993). The Earth Summit in Rio de Janeiro in 1992 brought together the heads of 120 nations to discuss global environmental issues (Soros 1994); agreements on climate and biodiversity represented a new step in global environmental awareness and action. If a full accounting could be made, Western European governments probably passed more environmental legislation in the second half of the 1980s than in any other five-year period.

Beyond these indications of the growing importance of the environmental issue, this study has focused on the institutional and political development of the environmental movement. Beginning with a modest organizational base in the 1960s, the movement has grown dramatically in scale and complexity. A variety of old and new conservation groups now tap public interest in wildlife protection and the preservation of nature and mobilize public support for these causes. In addition, a new wave of ecology groups has expanded the scope of the movement to include the emerging environmental problems of advanced industrial societies. In the process these ecology groups also critique the socioeconomic structures that create these problems.

We see environmental interest groups as central to the study of the larger environmental movement. Environmental groups are the primary policy instruments of the movement, the representatives of millions of citizens who are concerned about green issues. In virtually all European nations these groups are highly visible participants in the policy process. They are the organizations that testify before government committees, organize public protests, and pressure political parties on environmental matters. These groups are more than just representatives of the public, however. Through their decisions and actions they help to define the meaning of environmentalism. A group's choice of issues, decisions on how to frame these issues for the public, and choice of political tactics create an identity for a group. When summed across all groups, these choices give a meaning to the environmental movement as a whole. We therefore maintain that environmental interest groups provide a key to understanding both the nature of the green movement and the broader political changes occurring in advanced industrial democracies.

The Green Rainbow

One recurring theme in our study has been the organizational diversity of the environmental movement. At the local level one finds a rich organizational life through ad hoc citizen groups and ongoing initiatives (Donati 1984;

Diani 1990a; Gundelach 1984). An accounting by the European Community in the early 1980s identified several thousand local groups across Europe. More recently, green political parties have created a parliamentary arm of the movement. At the international level, there are formal coordinating bodies, such as the European Environmental Bureau and the International Union for the Conservation of Nature, as well as extensive networks of international conferences and policy forums.

We have looked at only a subset of this movement—national organizations concerned with environmental issues—but even within this stratum the diversity of organizational life is striking. The most prominent organizations are the mass-membership groups that represent and mobilize individual citizens. National federations, another important part of the movement, coordinate the actions of separate national interest groups or serve as coordinators for local environmental groups. The movement also incorporates a variety of environmental research institutes and educational foundations. Finally, several small groups of environmental elites facilitate communication among social and political leaders who are interested in environmental matters. Formal elite groups exist at both national and international levels (e.g., Ecoeuropa) and are joined by extensive informal networks of interaction.

Among mass-membership groups the diversity of organizational characteristics is tremendous.[1] Groups obviously differ in size, ranging from large well-funded groups, such as WWF or the RSPB, to smaller and lesser-known organizations such as the British Conservation Society or the Dutch Foundation for Alternative Life. Such groups also vary in internal organizational structure. At one extreme, one sees organizations in which members are simply passive supporters of the group, with no formal role in defining the organization's activities. FoE illustrates a contrasting pattern, however, one that stresses participation and attempts to involve the membership through local chapters and FoE-sponsored activities. These patterns of membership relations often overlap, creating great variation in decision-making structures within mass-membership groups. Most groups have centralized and bureaucratized methods of setting broad policy goals, though a few emphasize decentralized and participatory decision making.

The diversity of environmentalism is equally apparent in the political activities of these groups. Popular accounts and prior scholarship accentuate the unconventional nature of environmental action, represented by mass protests, spectacular actions, and dramatic events. This is an important part of the movement's political repertoire, but we also find extensive involvement in a wide array of conventional actions: meetings with government officials, participation in government commissions, meetings with MPs and parliamentary committees, and similar activities. Furthermore, these activities are unevenly distributed across environmental groups; some stress direct political action, whereas others have nearly accommodated themselves to a neocorporatist style of political involvement.

The most important source of diversity, however, is the ideological variation that exists within the movement. To be green can mean many things. For some environmentalists, whom we label conservationists, being green means preserving species and protecting the natural environment; their modest criticisms of society focus on excesses of growth or development that upset the natural balance. For others, namely, ecologists, greenness includes a concern with problems of nature and humankind that are seen as inevitable consequences of economic relations and life-styles in advanced industrial societies. Ecologists focus on a different set of issues, which they evaluate as part of a broader critique of the existing social and political orders. Between these ideal types of conservationist and ecologist are many shades of green thinking.

The environmental groups in our survey represent all these variations in the green rainbow. If each of our groups represents a viable form of environmental action, then there is no single model of what an environmental group is (or should be).

Because of the norms of centralization and structure that characterize contemporary politics (and resource mobilization theories of social movements), the diversity of the environmental movement (and other New Social Movements) is often seen as a sign of weakness. Lynton Caldwell (1990, 89–97), for example, maintains that this fragmentation prevents the movement from presenting a united front to policy makers and to potential allies and opponents, thereby impeding its policy impact. The logic of RM theory advances a similar view—that fragmentation among contemporary environmental groups produces inefficiency in mobilizing resources, a duplication of effort, and therefore a muted political impact. Several of our respondents shared these concerns, as they worried about competition over resources and influence with other environmental groups.

Although we do not disagree with these arguments, we think it is more appropriate to view environmentalism as a movement for social change, and not just as a traditional-style interest group representing a new set of issues within the governmental arena. Luther Gerlach and Virginia Hine (1970) have been the most prominent supporters of this opposing view, arguing that diversity is actually a source of strength for political movements. They maintain that the most effective movements for social change are those composed of a diverse array of groups, with many competing leaders and political bases linked through formal and informal networks (*SPIN* networks in their terms). This pattern is an apt description of the environmental movement not only in the individual nations we studied but in Western Europe as a whole.

The advantages of a diverse social movement appear in several areas. In terms of mobilizing resources, for example, different interests and political styles enable environmental groups to attract a wider array of supporters than would be possible with but a single organizational point of view. Some people are interested in wildlife conservation and preservation of the natural environment; others are more concerned with acid rain and toxic wastes. Different groups emerge to represent all of these interests. Equally important, people vary in their orientations toward social change. Some support campaigning groups that challenge the political establishment; others endorse more modest objectives. Individuals also vary in their degree of involvement. Some are drawn into groups like FoE that have active local chapters; most, however, pay their annual membership fee and are otherwise inactive.

These considerations also apply to institutional contributors and other sources of organizational support. A single environmental group might be more successful by adopting a large centralized structure, as William Gamson and others have shown (Gamson 1975; McCarthy and Zald 1977). Yet a diverse offering of "products" can maximize the resource pool of the movement. In the effort to save whales, WWF can mobilize the support of social elites ranging from the Agnelli family to the House of Windsor, whereas Greenpeace can marshal the support of students and critics of the social establishment. A unified environmental movement might create a firmer foundation for the movement, but one with a narrower social base.[2]

Similarly, diversity in appeal and style can broaden the political network of the movement, enabling different groups to develop alliances with a wide network of other interest groups and political institutions. Many of the social groups that are willing to work with the RSPB would never consider supporting Greenpeace, and vice versa. In chapter 7 we recounted how one conservation group approached members of the social elite to support a pending legislative proposal. At the same time, an ecology group was mobilizing support for the same proposal through its distinctly different networks. Neither group could have recruited the allies of the other; but by acting separately the movement was able to expand alliances.

Diversity, in terms of flexibility, may also be a tactical advantage in pressing the government for political reform (Gerlach 1971). FoE can do effectively what a bird society cannot. For instance, representatives of several wildlife groups cited the value of a protest or spectacular action by another group in focusing public attention on a pressing issue. Yet they admitted being constrained from using such tactics for fear of alienating their conservative membership. If we assume that to be successful a movement must vary its tactics with respect to the issue, opportunity structures, and other factors (Turner 1970), then a variety of groups gives the movement access to a broad range of political tactics.

From the policy side, the diversity of SMOs means that change is being pursued through several channels. In addition to petitioning the ministry of the environment, an environmental group may also be lobbying local governments, meeting with members of parliament, bringing petitions in the courts, contacting political parties, and appealing directly to the public. For a government agency trying to resist pressure, the environmental movement must resemble a Hydra that threatens from all sides simultaneously.[3]

Such diversity can be a positive attribute, especially during the rapid evolutionary period of a social movement, when different SMOs are able to explore numerous facets of environmentalism. While some groups are exploring sustainability and its implications, others are focusing on questions of bioethics or eco-feminism. This joint exploration of what it means to be green helps define the larger parameters of the movement and identify the areas of political change that are likely to reap the greatest rewards (for the movement and humankind).

Furthermore, competition among SMOs may keep the movement from becoming complacent and encourage a continued commitment to its original political goals. Social movements continually face the tension of balancing their calls for fundamental change against the attractions of pragmatic reform. The appeal of working within the system and yielding on small issues can sometimes deflect a movement from its ultimate goals of social change.[4] Organizational and ideological diversity within the movement can check these moderating tendencies. For instance, when Josef Leinen, former head of BBU, joined the political establishment by becoming a minister in a state government, the political role of the BBU was weakened. Other ecology groups, such as Robinwood and Greenpeace, nonetheless continued to press an ecologist orientation. This transformational process is clearly apparent in the history of the environmental movement. When the Sierra Club became too constraining, one of its founders established yet another organization, Friends of the Earth. The new green groups that seem to be forming continually challenge other groups to maintain their commitment to the fundamental values of the movement.

There is one caveat, however, to the beneficial aspects of diversity. Autonomous action might not generate the benefits we have described if groups pursue conflicting strategies or run at cross-purposes. To ensure that political action is cumulative, some coordination of action must take place, even if the parameters are informal and broad (Gerlach and Hine 1970). As we have shown, a variety of personal and political ties serve as communications channels within the environmental movement. The flow of personnel within the movement provides a shared experience and a personal link between groups. Informal discussion and coordination among environmental groups, though confirmed only indirectly by our data, are extensive (see also Diani 1990b, 1994). In most nations formal structures of discussion and interchange also exist, through environmental federations or participation on government advisory commissions. For instance, the European Community provides funding for the annual Brussels meeting of the groups in the European Environmental Bureau so that EC officials can meet with group representatives. This meeting also gives European groups an opportunity to exchange information, discuss their common interests, and plan their attempts to influence the Community—with the EC paying the bill. To be successful a movement must be a network of networks, and the contemporary green movement follows this pattern.

In short, the political logic of challenging social movements differs from that represented in the traditional hierarchical model of economic interest groups. As with biological diversity, political diversity can be a source of strength.

Ideologically Structured Action

We searched for factors that might explain the diversity among environmental interest groups. Resource mobilization provides one framework: this quasi-economic model of organizational behavior holds that smos act as maximizing organizations in pursuit of their objectives. Indeed, environmental groups are motivated to acquire resources, find strong allies, and pursue effective political tactics. RM theory would explain diversity as a function of the resource needs and opportunity structures facing smo as they pursue these goals. The structure and function of an organization is one such factor (Zald and Ash 1966); educational institutes, for example, have different resource needs than large mass-membership groups. In addition, variation in resources presumably reflects the effectiveness of each organization's ability to compete in the political marketplace. Efficient groups with popular goals should be more successful in amassing resources and political influence than disorganized groups without a clear political objective. RM theory essentially posits that smos are rational, value-free actors that pursue whatever course will maximize the organization's goals.

Throughout this study, however, we have found that environmental groups often act in ways that contradict RM theory. Environmental groups, it would appear, do not all share the same goals. Or if they do, they apparently evaluate their maximal options for political influence differently.

In chapter 1, we introduced the framework of Ideologically Structured Action to explain the behavior of environmental interest groups. In the case of smos particularly, behavior is not a value-free process in that the distinctive values of the movement justify its existence. Instead, the actions of smos, and probably most political groups, are influenced by political beliefs (or ideologies) that interact with resource needs and opportunity structures. A group's political identity tells it which goals are most valued, which problems need to be solved, which methods are appropriate, and what to expect from the government and other political actors. For example, although Greenpeace and the German Federation for Bird Protection are both concerned with protecting wildlife, their differences in approach to this problem can be traced back to their contrasting definitions of environmentalism and methods of reform. Moreover, a group's orientations affect the resource options and opportunity structures available to it. Greenpeace and the Federation for Bird Protection draw upon different resources bases and exercise distinct means of political influence. These ideological factors also underlie, we believe, the arguments that environmental groups and other nsms differ from traditional models of interest group politics and the expectations of RM theory. To study a social movement without incorporating its ideology into the analysis is like trying to read a book without opening the cover.

Although participants in the environmental movement share greenness and broadly similar political concerns, differences in ideological orientation ultimately become part of a group's identity. Conservation groups are interested in wildlife preservation and nature protection and pursue these issues based on an acceptance of the prevailing socioeconomic order. Ecology groups, in contrast, are concerned with modern environmental issues such as industrial pollution, nuclear power, and preserving the quality of life. More important, ecologists see these problems as emanating from the socioeconomic structure of advanced industrial societies. Thus, their interests reflect an adherence to the New Environmental Paradigm that represents a basic critique of the established order (Milbrath 1984, 1989; Cotgrove 1982; Dunlap and van Liere 1977).

The dichotomy between conservationists and ecologists is admittedly simple, but it is a valuable tool in explaining the behavior of environmental interest groups. New Social Movement researchers have focused on ecologists as the vanguard for a new society. Ecologists' commitment to an alternative social paradigm should lead them to adopt a range of organizational and political behaviors that distinguish them from conservation groups—even though both share a broad concern for environmental reform. The framework of ISA should explain the contrasts that flow from the different political orientations of these two types of groups.

The process of resource acquisition reflects the value of incorporating political orientations into the study of SMOs (see chap. 4). As RM theory would suggest, virtually all environmental interest groups recognize the need to acquire resources to support the organization and its goals. The framework of ISA focuses more specifically on the nature of those resources. Conservation groups emphasize issues and values that are widely shared by most Europeans; consequently, these groups are able to assemble large memberships, large budgets, and large staffs. More important, conservation groups are able to solicit money from members of the political establishment: government agencies, business associations, and foundations. Ecology groups, however, by virtue of their challenging political orientation, must often find support outside the political establishment, through membership fees, the sale of group materials, and individual gifts. It seems obvious that important differences should exist between the potential funding of a national bird society, for example, and the funding available to a radical "deep ecology" or anti–nuclear power group—but this means that resource acquisition is not a value-free process that is interchangeable across SMOs.

The role of ideology in creating the context for political action is further reflected in a group's perception of the world of environmental politics. Following David Snow and Robert Benford's (1988) concept of diagnostic framing, we find that a conservationist or ecologist orientation frames a group's identification of key issues, the actions necessary to address them, possible obstacles to reform, and potential allies and opponents (chaps. 6 and 7). In creating these perceptions of the political world, whether they are accurate or not, ideology defines the context for environmental action. Sometimes these perceptions are well founded. Ecologists do not just imagine that businesses are opposed to their goals; business interests *are* opposed. Thus, ideology partially creates the opportunity structures facing an SMO.

The environmental identity of a group ultimately culminates in its choice of political strategies (chaps. 8 and 9). The options available to a challenging group, for example, differ from those of a socially accepted organization, such as Civic Trust or the RSPB. Because of their identity and their closer ties with the political establishment, conservation groups are more likely to engage in conventional political activities such as sitting on government commissions, meeting with government ministers, and meeting with members of parliament. Ecology groups do so as well—in fact, to a greater degree than the prior literature would suggest—yet they balance these activities with an almost equal level of unconventional political action: protests, spectacular maneuvers, public demonstrations, and the like. Our discussions with environmentalists made it clear that they selected tactics not just for efficacy but also because of compatibility with the identity and goals of the group. The identity of ecology groups is linked to their status as challengers to the political establishment, and the best way to illustrate that identity is through protest activities.

To establish the centrality of environmental orientations in determining behavior, we entered the following characteristics of the groups into a principal components analysis: environmental orientations, resource base, organizational structure, evaluations of other actors, and political tactics.

The first dimension from this analysis clearly illustrates the importance of ideological orientations in structuring the organizational and political characteristics of environmental groups (table 10.1).[5] The two highest loading variables on the dimension are ideological: the ecologist or conservationist orientation of a group (.85) and its Left/Right placement (.84). Protest activities also have a strong positive loading (.66); conversely, there are strong negative loadings for participation in conventional channels, such as government commissions ($-.52$) and formal meetings with ministry officials ($-.45$). As we saw in chapter 4, environmental orientations are related to the resource base of environmental groups as represented by membership size, professional staff, and annual budgets. Ecology groups generally have fewer resources than conservation groups. Finally, perceptions of other political actors are also related to this dimension. Ratings of green parties are higher among ecology groups; ratings of establishment actors, such as the prime minister, conservative parties, or business associations, are higher among conservation groups. In summary, environmental orientations emerge as a central component in structuring the political perceptions and political actions of environmental interest groups.

The principal components analysis also highlights several characteristics that are unrelated to the ecologist-conservationist continuum. Although participatory norms are a prominent theme in the ideology of NSMs, ecology groups do not have a distinctly participatory structure. Most environmental interest groups have adopted a centralized structure controlled by the leadership (chap. 4), and ecology groups deviate only slightly from these orga-

TABLE 10.1

A Principal Components Analysis of Group Characteristics

Ecological orientation	.85
Leftist ideology	.84
Protest activity	.66
Age of group	−.60
Government commissions activity	−.52
Large membership	−.47
Large professional staff	−.47
PM environmental rating	−.46
Ministerial activity	−.45
Decentralized decision making	.39
Business environmental rating	−.35
Conservative party rating	−.29
Green party rating	.20
Membership input	.19
Media activity	.11
Party activity	.03
Eigenvalue	3.77
Variance explained	23.6%

Figures are factor loadings on the first dimension of an unrotated principal components solution. Pairwise deletion of missing data was used.

nizational tendencies. Decentralized decision making (.39) and opportunities for membership input (.19) are only modestly related to this dimension. National environmental groups represent a difficult setting in which to find the participatory values of NSMs; such norms are more likely to flourish in local organizations or ad hoc groups. Within national interest groups, the impact of ideology is not sufficiently strong to overcome the logic of centralization and hierarchy.[6]

Two important political activities—contact with the media and with political parties—are unrelated to the ecologist-conservationist dimension. Media visibility is so vital to public interest groups that is equally important to both ecologists and conservationists. When 86 percent of all groups say they are often in contact with the media, there is little opportunity for variation. In contrast, partisan contact is the least frequent activity for environmental groups. In spite of the proliferation of green parties and attempts by many established parties to cater to environmental issues, neither ecology nor conservation groups see partisan activities as a prominent part of their political repertoire (chap. 9). Because of the prominence of political parties in the policy process of most European states, this rejection of partisan politics may represent an important stylistic change of the environmental movement. Environmentalists generally reject the intermediary role of parties and instead seek to pressure policy makers directly or through the force of public opinion. If environmentalists are successful in avoiding the aggregating role of political parties, their stand may encourage other interests to pursue similar strategies. Innovative efforts to change the structure of interest representation, such as greater use of referendums and citizen advisory groups, may contribute to a decline in the functional role of political parties (Dalton et al. 1984, chap. 15).

When viewed overall, ecology groups reflect an important clustering of characteristics that are also prominent in the scholarship on NSMs, even if the overlap is partial or incomplete. First, this set of characteristics underscores the argument that NSMs represent a different pattern of politics, one that involves ideological orientation, perceptions of the political world, political alliances, and political tactics. Second, it can reinforce key features of the environmental movement, especially the challenging orientation of ecology groups. Charles Tilly, for example, suggested that SMOs that avoid the institutionalization and professionalization inherent in organizations are better able to resist the transformation of goals from social change to organizational maintenance (Tilly 1978, 58–59). At a more practical level, Robert Hunter (1979, 247) noted that the founders of Greenpeace saw organizational and strategic choices as interrelated (see also Barkan 1979; Downey 1986). The interaction among group characteristics thus both reflects and reinforces the identity of a challenging social movement. This overlap strengthens a movement's ability to retain its political values without accommodating itself to the political status quo.

Third, the importance of ideology in structuring environmental action also yields a basic critique of a rational-choice approach to the behavior of political organizations. Political organizations, especially SMOs, are not value-free actors that simply maximize short-term payoffs. In pursuit of its political goals, for instance, an ecology group accepts constraints on resources and political opportunities that a conservation group does not have to face.[7] This situation is not the maximization strategy implied by RM theory, because if an ecology group really wanted to increase its membership and gain greater political access, it would shed the ideology and political style that define it as ecologist. One might argue that a group attempts to maximize its success within the political framework of its identity. This, however, relegates the maximizing logic of rational choice to a near tautology: tell us what a group wants, and RM can predict that it will try to satisfy these objectives. In such a case, the more powerful theoretical statement is to define the political framework within which decisions are made.

The importance of organizational ideology suggests the need for more research. We have contrasted two shades of green—conservationist and ecologist—but more refined ideological categories are required (see, e.g., Paehlke 1988; Bramwell 1989; Eckersley 1992). The conservation-ecology dichotomy is a potent predictor of organizational behavior, but it is still imperfect. The principal components analysis links group characteristics to environmental orientation, but some ecology groups depart from this pattern. Greenpeace, for example, criticizes society while maintaining a consciously undemocratic organizational structure; for this group, criticism of Western society is not paired with principles of democratic participation. Other ecology groups, such as many chapters of FoE, successfully combine these two features. Similarly, some ecology groups press for fundamental changes in European economic systems and living styles but are unwilling to exceed the bounds of conventional politics in pursuit of that cause.

We have generally dealt with ideology in a static way, without exploring its origins or how it develops over time. As we noted in earlier chapters, environmental beliefs are still evolving, and there is no clearly defined ideology of environmentalism that could match the prescriptive model of the socialist labor movement, for example. The evolution of the environmental movement may therefore provide a unique opportunity for social scientists to study the development of a new political ideology. The dynamic nature of environmentalism also provides a laboratory for studying political change. Some environmental groups, for example, the American Audubon Society, have shifted their emphasis from conservation to ecology. Examining how group identities are formed and reformed is vital to understanding the full dynamics of Ideologically Structured Action. We strongly agree with Alberto Melucci's (1989) claim that the formation of group identities should be a central research question in the study of NSMs. Further refinements in defining and explaining the ideology of SMOs can only increase the value of this approach.

The Green Movement and Democratic Politics

If we step back from our analyses of environmental interest groups and the variations in goals and style existing within the movement, our findings have wide-ranging implications for politics in advanced industrial democracies.

In this study, we have identified the green movement as an important new development for Western political systems, as warranted by its size alone. The significance of environmentalism goes further, however. More than just new actors on the political stage, the green movement and other NSMs typify the broader political changes that are transforming the political process in advanced industrial democracies. If green politics is an indicator of future political developments, then what might this future hold?

As we argued elsewhere (Dalton and Kuechler 1990, chap. 14), NSMs represent a definite, though not revolutionary, challenge to the established political order of Western democracies. Environmentalists do not want to destroy the capitalist economic system or to overturn the democratic political process. Radical and anarchic elements within the movement are scarce, even if they are visible in the media. Rather, most environmentalists are products of Western affluence and strong advocates for the democratic creed. The green movement, and other NSMs, primarily represent a reformist challenge to contemporary political systems. They press for political change across a wide front, often asking simply that European societies continue the political and social thinking that initially spawned these movements.

Environmentalism represents, above all, an expansion of the political agenda to include noneconomic issues as central political concerns. Repeated public opinion surveys document the substantial concern of Western publics for environmental issues. National and local governments are now incorporating environmental considerations into their policy making and passing on a growing framework of environmental regulations to business, industry, and the consumer. Furthermore, green issues are joining other quality-of-life concerns: issues of women's rights, consumer protection, and life-style choices. We see these new interests not as supplanting the traditional economic interests of Western democracies but as adding a new political dimension that competes with and potentially contradicts established lines of political division. Environmentalism epitomizes these new concerns.

The environmental movement also represents a change in the style of European interest group politics. Most European unions, business organizations, and other economic interest associations follow the neocorporatist model of unifying diverse interests under an umbrella group (Berger 1981; Grant 1985). The green movement, however, has adopted a pattern of organizational diversity, whereby interest groups compete for membership support and influence within the political process. Even when umbrella groups or federations exist within the environmental movement, they function more as coordinating and networking systems than as authoritative voices for the movement.

The political style of the environmental movement also differs from the normal patterns of conventional lobbying activities. Some environmental groups have established close working relations with government agencies and parliamentary committees. Yet unconventional behavior constitutes a significant part of the political repertoire of some green groups, though it is not as common as some analysts suggest. The green movement therefore combines the protest politics of the 1960s and 1970s with the pragmatic use of conventional methods.[8] Some groups may moderate their tactics with time and with their integration into the polity—even the RSPB was once known for its disruptive political tactics—but it is unlikely that the environmental movement will completely shed its unconventional political style. Even if some green groups do become more conventional, new challenging groups will take their place. The open structure of the environmental movement means there is no monopoly of representation as there is under neocorporatist systems. Furthermore, direct-action tactics and spectacular actions arise from both the political identity of the movement and the need to mobilize support and communicate with a dispersed and loosely integrated public base. The intermixing of strategies therefore represents another characteristic of politics that we can identify with the development of New Social Movements.

Many of these features of environmentalism look different for analysts of American politics than they do for analysts of European politics. Public interest groups and political action committees are a regular part of the American political process, and the loose structure and style of such groups reflect the pluralist patterns of American politics. The situation is different in Western Europe, however, where New Social Movements are more politicized and ideological. The United States, for instance, has no equivalent to the green political parties that have formed throughout Europe, and even American ecology groups appear less ideological than their European counterparts. Equally important, the open style and decentralized structure of American politics enables the American environmental movement to integrate itself into the political process. In comparison, European environmental groups are competing in a world of neocorporatist economic interests and closed decision-making structures, without freedom-of-information laws and often without traditions of direct citizen involvement in the policy-making process. Because of these constraints, NSMs represent a fundamental political challenge to the structured and formalized political processes of most European states.

If the European green movement is politically successful, it may at least partially reshape the style and structure of democratic political processes in these nations. The growth of environmentalism represents a shift from centralized patterns of decision making to a dispersal of political authority. The new contenders do not speak with a single voice or pursue social change through a single program, which means that policy makers must deal with a fluid and competitive pattern of interests. In addition, the arenas for decision making are likely to spread—to local governments, the courts, national parliaments, and international bodies. The diversity of local environmental initiatives illustrates that policies enacted at the national level cannot simply be implemented at the local level. The diversity of NSMs will challenge the pattern of regularity and predictability that is identified with European interest group politics.

These changes in political style can lead to institutional reforms in the policy process. Environmental groups have pressed for more direct citizen participation in policy making. In Germany, for example, local groups have won changes in German administrative law to allow for citizen participation in local administrative processes (Hager 1993). In Italy, environmental legislation now grants individuals legal standing in the courts when they seek to protect the environment from the actions of municipalities or government administrative agencies; and environmental groups have expanded the use of referendums to involve the public directly in policy making. The Dutch national debate on nuclear power reshaped the way government deals with public interest policies (Jamison et al. 1990, chap. 4). Because such institutional changes are difficult to accomplish, they will likely precede at a slower pace than policy change; but once implemented, they will restructure the whole process of policy making to extend beyond a single issue or a single policy agenda.

Thus, the environmental movement has become an instrument for social and political change, although the extent remains uncertain. The diversity of the movement accounts for part of this uncertainty. As we have argued, the objectives of conservation groups call for only modest social reform and do not challenge the basic structure and paradigm of Western industrial societies. The accomplishments of the green movement already meet the prime objectives of conservationists in many areas. Ecologists, in contrast, are agents for a more fundamental social change that would transform life-styles, economic relations, and political structures. If ecologists are correct in claiming that only dramatic changes such as these will save the planet from environmental crisis, the achievement of their goals becomes essential. Yet, many of our findings suggest that ecologists are being drawn away from their transformational goals of reordering society toward more immediate and pragmatic policy objectives.

Ecology and conservation groups must consider what is feasible in addition to what is ideal if they wish to make progress. To satisfy their constituencies, mass-membership groups also must be able to point to immediate results. Saving the harp seal will not transform the world, though it may be a small step in the right direction (in addition to generating sympathetic press coverage and donations). As William Gamson has argued (1975), there is a logic to "thinking small." This tension between pragmatism and fundamentalism exists within the green movement and may lead to a moderation of its demands for social change. Our findings suggest, however, that even ecologists feel the need to balance these two contending forces, as is reflected in their political actions.

The beliefs and behaviors of ecologists already represent a reformist orientation toward society, which allows transformational goals to be delayed or moderated. Accommodation is made easier by the broad, but unspecific, nature of the alternative paradigm of environmentalists and other NSMs. A great deal has been written about our current environmental problems, but the alternative model of a green future is still widely debated within the movement and by society at large. The environmental identities we have stressed throughout this book are still evolving. There is a movement for social change, but no clear utopian vision of what the future should look like.

Given this evidence, we think it unlikely that ecologists will be agents for rapid and radical social change. Instead, we shall probably witness a continuation of the incremental progress of the past two decades. Rather than supplanting the dominant social paradigm, ecologists will modify it. Rather than transforming political systems, they will reform them. Nonetheless, the developments we have already seen suggest that the potential for substantial and meaningful change is real. Whether these accumulated steps will produce the transformation that some environmentalists deem necessary for the health of the planet remains to be seen.

Appendix

A C O M P E N D I U M   O F

E N V I R O N M E N T A L   G R O U P S

This study examines the political orientations, policy activities, and structural characteristics of the major environmental organizations in Western Europe. We defined this charter quite broadly to include the diverse groups concerned with environmental issues so that we might compare different elements of the environmental movement—that is, different shades of green. In this appendix we describe the data collection procedures used in this study and present basic descriptive information on the major environmental groups in Western Europe.

The first task we faced was to identify the population of relevant organizations. The European Environmental Bureau (EEB) and two of its directors, Ernst Klatte and Hubert David, provided invaluable assistance in this task. The EEB is an international association of national environmental groups that was formed in 1974 (Lowe and Goyder 1983, chap. 10). For the member groups, it provides a communication channel with the European Community and a permanent lobbying presence in Brussels. For the Community, the EEB serves as a consultative body on environmental issues. Its financial support comes from the European Communities, national governments and foundations, and membership fees.

Our enumeration of environmental organizations began with the fifty-seven groups that were members or associate members of the EEB in 1985 (we eventually interviewed forty-six EEB members). Ernst Klatte provided information on each of these groups and invited us to attend the annual Brussels conference of the bureau members. The EEB membership includes most of the large environmental groups in Western Europe. If there is any bias in the EEB membership, it is the underrepresentation of small, politically assertive, and unconventional environmental groups. The EEB also excludes environmental groups that are explicitly linked to a political party. We therefore supplemented the EEB list by including the European affiliates of Friends of the Earth and Greenpeace that were not already EEB members. All of the European affiliates of World Wildlife Fund were also added to our master list at this point. Through literature searches and consultations with numerous colleagues, we expanded this list by the addition of the smaller or less conventional groups that were not EEB members, such as Robinwood in Germany, Lega per l'Ambiente–Arci in Italy, and NOAH in Denmark. These procedures yielded a master list of approximately ninety organizations, which served as the initial base of our study.

During the summer and fall of 1985, we contacted each group on the master list and requested copies of their newsletter for members, recent policy papers, and annual report. The vast majority furnished the requested materials, which provided background information on the structure of the groups and a chronicle of their activities over the previous year. In sending out these requests, we also identified several organizations that had ceased to exist (for instance, Friends of the Earth in Ireland and Germany) and thereby slightly shortened the master list.

Over these same months the research team in Tallahassee, Florida, developed a standardized questionnaire representing our research interests. To provide an opportunity for comparative analysis, portions of this questionnaire were drawn from Lowe and Goyder's (1983) study of British environmental groups and Frank Wilson's (1987) study of French interest groups; other items were drawn from the Eurobarometer studies of public attitudes toward the environment that paralleled our survey. The questionnaire was pretested on several environmental groups in Tallahassee, revised, and then translated into French, German, and Italian.

After reviewing our master list, we arranged to interview representatives of the organizations that we felt were major environmental actors in their nation or that represented an important element of the overall environmental movement. In preparation for our interviews, we narrowed the list by excluding regional affiliates of national organizations; for instance, we retained the Council for the Protection of Rural England but not the Council for the Protection of Rural Wales.

In November 1985 our research team traveled to Europe to interview group representatives; the final interviews were conducted in spring 1986. Using the procedure common to other interest group studies (e.g., Milbrath 1963; Berry 1977; Lowe and Goyder 1983; Schlotzman and Tierney 1986), a single informed respondent was interviewed for each organization. In two-thirds of the cases this respondent was the director of the organization; in most other cases we interviewed the deputy director or public information officer. Although questions of the perceptual accuracy of the selected respondent are unavoidable in research of this type, we feel these problems are not substantial in our study. First, we chose to study small, relatively homogeneous organizations and not the large complex structures that are typical of studies on organizational behavior (for example, political parties, businesses, or large economic associations). The modal environmental group had only six full-time employees; individual respondents were normally well informed about the organization, especially if he or she was head of the group. Second, we were sensitive to the potential problem of projective responses in designing the questionnaire. We limited subjective questions and focused on factual aspects of the groups: structural characteristics, the issues the group had pursued over the previous year, and the tactics that were used. Third, where possible we utilized external checks on the interview material. In fact, given our initial concerns about access to the groups and the frankness of respondents, we were struck by the exceptional receptivity shown us in almost every case. Respondents appeared willing to speak frankly and confidentially partially because we were members of an international research team and not involved in the politics of their nation.

The typical interview lasted approximately ninety minutes, based on a common structured questionnaire adapted to the national context (a copy of the questionnaire is available from the author). Various members of the research team spoke French, Greek, Italian, or German (in addition to English), and when possible interviews were conducted in the respondent's native language. Several organizations could not be contacted for logistical reasons, and only one refused to be interviewed.

We surveyed a total of sixty-nine environmental organizations, including nearly all of the major environmental organizations in Western Europe. The following table provides basic information on these organizations, as well as on several other significant groups that were not included in our study (noted by an asterisk). We include the type of organization, its membership base, and an estimate of its 1985 operating budget. When combined, these statistics provide impressive evidence of the magnitude of the contemporary environmental movement: these organizations boast a combined membership of nearly 10 million Europeans and an operating budget of over $50 million for 1985.

TABLE A.1

Major European Environmental Groups, by Country

Group	Year Founded	Group Type	Membership	Approximate 1985 Budget (in thousands of dollars)
Belgium				
Les Amis de la Terre	1976	Mass membership	1,000 members	14
Bond Beter Leefmilieu	1974	Federation	56 groups	97
Greenpeace	1979	Mass membership	6,000 members	116
Inter-environment Wallonie	1974	Federation	74 groups	233
National Union for Conservation	1952	Federation	29 groups	6
*Natuur 2000				
Raad Leefmilieu te Brussel	1973	Federation	18 groups	23
Réserves Ornithologiques	1951	Mass membership	18,000 members	194
Stichting Leefmilieu	1970	Educational foundation	No members	68
*Wielewaal	—	Mass membership	—	—
World Wildlife Fund	1965	Mass membership	25,000 members	466
Denmark				
Danmarks Naturfrednings Forening (DN)	1911	Mass membership	220,000 members	2,376
Dansk Ornitologisk Forening (DOF)	1906	Mass membership	8,500 members	196
Friluftradet	1942	Federation	72 groups	871
GENDAN	1974	Corporation	No members	351
Greenpeace	1980	Mass membership	45,000 members	1,098
NOAH	1969	Federation/members	50 branches/ 2,000 members	55
*World Wildlife Fund	1965	Mass membership	50,000 members	—
France				
Association des Journalistes et Ecrivains pour la Nature et l'Ecologie	1969	Elite membership	200+ members	10
Les Amis de la Terre (AdlT)	1970	Federation/members	120 branches/ 3,000 members	52
Comité Législatif d'Information Ecologique (COLINE)	1979	Elite members	100+ members	39
Comité Régional d'Etude pour la Protection et l'Aménagement de la Nature (CREPAN)	1968	Federation	19 groups	32

(continued)

Group	Year Founded	Group Type	Membership	Approximate 1985 Budget (in thousands of dollars)
Fédération Française des Sociétés de Protection de la Nature (FFSPN)	1968	Federation	100+ groups/ 850,000 members	519
Greenpeace	1977	Mass membership	7,000 members	454
Institut pour une politique européene de l'environnement (IPEE)	1976	Research institute	No members	104
*Ligue pour la Protection des Oiseaux	1912	Mass membership	5,000 members	581
Nature et Progrès	1964	Mass membership	6,000 members	260
Société Nationale de Protection de la Nature (SNPN)	1854	Mass membership	7,500 members	970
World Wildlife Fund	1963	Mass membership	15,000 members	325
Great Britain				
Civic Trust	1957	Foundation	No members	457
Conservation Society	1966	Mass membership	5,000 members	39
Council for Environmental Conservation (CoEnCo)	1969	Federation	22 groups	97
Council for the Protection of Rural England (CPRE)	1926	Mass membership	32,000 members	373
*Council for the Protection of Rural Wales (CPRW)	1928	Mass membership	3,000 members	70
Fauna and Flora Preservation Society (FFPS)	1903	Mass membership	5,000 members	118
Friends of the Earth (FoE)	1970	Federation/members	200 branches/ 27,000 members	689
Green Alliance	1978	Elite membership	500+ members	35
Greenpeace	1977	Mass membership	60,000 members	1,408
*International Institute for Environmental Protection	1971	Research institute	No members	—
*National Trust	1895	Mass membership	1 million members	—
*Royal Society for Nature Conservation (RSNC)	1912	Mass membership	160,000 members	403
Royal Society for the Protection of Birds (RSPB)	1899	Mass membership	407,000 members	9,860
*Scottish Civic Trust	1967	Foundation	No members	115
*Socialist Environmental Resources Association (SERA)	1973	Mass membership	500 members	—

(continued)

Group	Year Founded	Group Type	Membership	Approximate 1985 Budget (in thousands of dollars)
*Soil Association	1946	Mass membership	5,000 members	113
Town and Country Planning Association (TCPA)	1899	Federation	1,000+ planning bodies	465
*World Wildlife Fund	1961	Mass membership	80,000 members	5,585
Greece				
Elliniki Etairia	1972	Mass membership	4,000 members	67
EREYA	1969	Elite membership	500 members	33
Friends of the Earth	1978	Mass membership	100 members	3
Friends of the Trees	1902	Mass membership	100+ members	20
Hellenic Society for the Protection of Nature (HSPN)	1951	Mass membership	1,000+ members	8
PAKOE	1979	Research institute	No members	133
Ireland				
An Taisce	1947	Mass membership	6,000 members	96
Wildlife Federation	1979	Mass membership	2,000 members	26
Italy				
Agriturist	1965	Mass membership	10,000 members	146
Amici della Terra (AdT)	1977	Federation/members	1,000 members	87
Fondo per l'Ambiente Italiano	1975	Mass membership	2,000+ members	350
Italia Nostra	1955	Mass membership	15,000 members	1,000
Lega per l'Ambiente–Arci	1980	Mass membership	30,000 members	584
Lega per l'Abolizione Caccia	1978	Mass membership	4,000 members	9
Lega Protezione di Uccelli (LIPU)	1965	Mass membership	23,000 members	229
*Federnatura	1959	Federation	No members	—
*Pro Natura	1948	Mass membership	20,000 members	—
World Wildlife Fund	1966	Mass membership	62,000 members	584
Luxembourg				
*Greenpeace	1980	Mass membership	—	—
Mouvement Ecologique	1981	Mass membership	2,000 members	87
Natura	1971	Federation	40 groups	29

(continued)

Group	Year Founded	Group Type	Membership	Approximate 1985 Budget (in thousands of dollars)
Netherlands				
Greenpeace	1978	Mass membership	80,000 members	1,071
Institut voor Natuurbeschermingseducatie (IVN)	1960	Educational institute	14,000 members	357
*De Kleine Arde	1972	Mass membership	13,000 members	—
Stichting Milieu-edukatie (SME)	1975	Educational institute	No members	232
Stichting Mondiaal Alternatief	1972	Mass membership	1,500 members	9
Stichting Natuur en Milieu (SNM)	1972	Federation	17 groups	71
Vereniging Milieudefensie (VMD)	1972	Mass membership	15,000 members	321
Vereniging tot Behoud van de Waddenzee	1965	Mass membership	31,000 members	715
*Vereniging tot Behoud van Natuurmonumenten	1905	Mass membership	250,000 members	11,928
Vereniging tot Bescherming van Vogels	1899	Mass membership	30,000 members	714
*World Wildlife Fund	1972	Mass membership	100,000 members	—
(West) Germany				
Bund für Umwelt- und Naturschutz Deutschland (BUND)	1975	Mass membership	28,000 members	1,000
Bund für Vogelschutz (DBV)	1899	Mass membership	140,000 members	400
*Bund Naturschutz Bayern	1913	Federation	60,000 members	2,770
Bundesverband Bürgerinitiativen Umweltschutz (BBU)	1971	Federation	350 groups/ 140,000 members in subgroups	200
Deutscher Naturschutzring (DNR)	1960	Federation	95 groups/3.3 million members	218
Greenpeace	1980	Mass membership	75,000 members	3,000
*Öko-Institut	1977	Research institute	4,000 members	428
Robinwood	1982	Mass membership	750 members	48
Schutzgemeinschaft Deutscher Wald	1947	Mass membership	22,000 members	200
World Wildlife Fund	1963	Mass membership	40,000 members	3,300

Note: Groups marked by an asterisk were contacted as part of this study but were not interviewed. A dash indicates that data were not available.

Notes

Chapter 1: Environmentalism and Social Movement Theory

1. See chapter 3 for evidence on the environmental attitudes of European publics.

2. A third approach has been to study citizen groups within the tradition of pluralist interest group politics (see Walker 1991; Berry 1977; Schlozman and Tierney 1986). Because this research generally does not focus explicitly on social movement organizations, and sometimes actually treats social movements as an anomaly to the theory, we give it less weight in the following analyses.

3. The RM and rational-choice perspectives differ sharply in their expectations concerning the basis of individual participation in political groups (e.g., Olson 1965 vs. Fireman and Gamson 1979). At the group level, however, they yield very similar predictions, and we shall therefore discuss them jointly.

4. Indeed, the history of the environmental movement (discussed in chap. 2) has shown that the joint presence of entrepreneurial initiative and resource opportunities were important factors in the formation of many environmental organizations. We also find, however, that culturally based fluctuations in popular sentiments affect the potential for new SMOs to form (see Brand 1990).

5. Rational-choice theorists also maintain that the ability of entrepreneurs to benefit individually from a social movement organization (through salary, career advancement, and the like) helps these organizations to overcome Mancur Olson's (1965) well-known "logic of collective action" (Salisbury 1969; Frohlich et al. 1971).

6. A similar dichotomy of styles is noted within the women's movement and peace movement (Freeman 1975; Ferree 1987; Gelb 1989; Rochon 1988).

7. See, for example, the evidence assembled in Dalton and Kuechler (1990) and Klandermans (1989).

8. The relevance of ideology to NSMs is perhaps most clearly illustrated in past research on the women's movement, where the different ideological definitions of feminism have affected the structure and tactics of various groups (see Ferree and Hess 1985; Freeman 1975; Gelb 1989).

9. Jack Walker makes this point for interest groups as a whole: "Once an organizational *recipe* [identity in our terms] is chosen by group leaders, many other aspects of the association's organization and functioning are also determined" (Walker 1991, 15).

10. In extreme cases, a singular focus on ideological goals might even overwhelm the pragmatic needs of an organization. See the discussion of this point in Terry Moe's insightful study of political organizations (1980, chap. 5, esp. 134ff).

11. Many discussions of NSMs define the identity of these groups in terms of two dimensions: goals and methods (Kitschelt and Hellemans 1990; Frankland and Schoonmaker 1992). Because we see methods as partially dependent on the goals of the group, we shall focus on the goals or values of a group as the key to defining its environmental orientation.

12. William Gamson has drawn attention to the special role of media in diffusing these images (Gamson 1992; Gamson and Wolfsfeld 1993).

13. Spain and Portugal joined the European Community in 1986.

14. I wish to thank Jacques-René Rabier for the opportunity to suggest questions for inclusion in Eurobarometer 25 and for allowing me to include items from the Eurobarometers in the environmental group survey.

15. In three cases the interviews were conducted by mail because the offices of these groups were far from our central research sites.

Chapter 2: The Evolution of Environmentalism

1. For a rich enumeration of the various philosophical and ideological precursors to the contemporary environmental movement see Anna Bramwell's (1989) discussion of the role of environmentalism in earlier political ideologies. I disagree, however, with her implicit linking of many of these historical experiences to the present environmental movement.

2. Almost every nation stakes some claim to being at the vanguard of the environmental movement. English authors note that Henry VIII passed a law to protect "wilde fowle" in 1534; French environmentalism began with water-quality legislation in 1669; the Germans claim that early legislation protecting the continental forests stimulated the Industrial Revolution by necessitating the development of iron and steel as building products; others such as Nicholson (1970, chap. 6) and Boardman (1981, chap. 1) trace the environmental movement back to the ancient Egyptians and Greeks.

3. The signatories were Switzerland, Germany, Austria-Hungary, Belgium, Spain, France, Greece, Luxembourg, Monaco, Portugal, and Sweden. Not included among the signatories, though they participated in the conferences, were Great Britain, the Netherlands, Italy, Russia, and Norway. See Hayden (1942, chap. 2) for additional details.

4. Similar forces existed on the other side of the Atlantic. The American Audubon Society was founded in 1886, and the Sierra Club in 1892; the Boone and

Crockett Club was established as a nature conservation lobby in 1887 (see Hays 1959; Hayden 1942).

5. The organization has evolved over the years, shifting its concern from wild-life in British colonial holdings to nature conservation issues in Britain. Instead of protecting African big game, its recent campaigns have aimed at protecting British hedgehogs, toads, and bats. It is now known as the Fauna and Flora Preservation Society (FFPS).

6. The one exception to this pattern is Germany, where the "coordination" policies of the Third Reich led to the outlawing of some conservation groups and the restructuring of others, such as the DBV and the German Federation for the Protection of the Homeland, as subsidiaries of the Nazi-controlled Reich League for Nationhood and Homeland (Dominick 1992, 102–05). In addition, the Nazis formed their own conservation organization, the People's Federation of Nature Protection (Volksbund Naturschutz); see Wey (1982, 138).

7. Dominick (1992, 122) reports that the DBV, for example, lost nearly half of its membership in the immediate postwar period.

8. The organization was originally called the International Union for the Protection of Nature but changed its name in the 1950s to reflect a broadening of its concern to conservation of natural resources and other environmental issues.

9. Even more than the first environmental wave at the turn of the century, this second wave was an international or even global development. Similar trends, perhaps of even greater magnitude, occurred in the United States over this same period (Hays 1987; Scheffer 1991; Dunlap and Mertig 1992), as well as in other industrial societies (Kamieniecki 1993).

10. For an insightful study of the general diffusion of social movement activity from the United States to Europe, see McAdam and Rucht (1993).

11. FoE was the licensed company name of the national office, which led to an initially awkward relationship with autonomous local groups using the FoE title. The national office resolved this situation by offering a "franchise" to local FoE organizers for a nominal fee. The present organizational structure of FoE-UK reflects the legacy of the dual national/local base. See chapter 4.

12. Other research on green activists suggests that we should classify movement organizations in terms of both their goals (reformist/revolutionary) and their methods (conventional/unconventional); see, for example, Kitschelt (1989); Frankland and Schoonmaker (1992). We see the goals of the movement as close to our definition of group identity and therefore emphasize this dimension in our definition. The question of methods seems, in our opinion, to be a consequence of a group's goals more than a definition of a group's identity.

13. It is admittedly difficult to construct a consensual definition of the ecological identity, because the values of the movement are not clearly articulated. Ecologists are often more successful in diagnosing the problems of advanced industrial societies than in formulating possible solutions. Claus Offe (1990), for instance, maintains that one of the distinctive features of NSMs is their lack of a well-defined ideological utopia similar to those of earlier reformist movements in Europe. What follows is a summary of the main features of the new ecologist perspective that are commonly cited by ecologists and the academic literature.

14. A more holistic and fundamentalist version of this orientation is known as "deep ecology," in which nature is valued independently of human needs (see Bookchin 1982; Devall 1988; Naess 1988). Deep Ecology is most closely associated with environmental groups such as Earth First! (see Manes 1990).

15. We shall see in chapter 4, however, that alternative societal goals do not necessarily translate into participatory norms within ecology groups. At least for our sample, these appear to be relatively independent dimensions of group characteristics.

16. The central distinction between conservation and ecology orientations represents an ideological cleavage within the environmental movement that is widely observed. Stephen Cotgrove (1982) emphasizes the distinction between conservationists and "new environmentalists" in his research on British environmental groups. Lester Milbrath's (1984) study of environmentalism in three nations uses the classification of "nature conservation" and "environmental reformer" groups. Dieter Rucht (1989) classifies German and French environmental groups as representing either a "nature conservation" or a "political ecologist" orientation; Vadrot (1978) uses a conceptually similar classification of French groups as "syndicalistes" or "libertaires." Donati and Lodi (1988) distinguish among the "conservationist," "environmentalist," and "political ecology" currents of the Italian environmental movement (see also Diani 1990).

Chapter 3: Europeans and Environmentalism

1. My favorite example of this treatment is contained in the American comic strip Bloom County, in which Opus sails to Antarctica to visit his mother. After pulling out from port in New Zealand, he notices that he is surrounded by long-haired, bean-curd-eating, whale-song-singing eccentrics. In the last frame of the comic strip, he realizes he is on the Rainbow Warrior surrounded by "environmental extremists."

2. These data are available from the Inter-university Consortium for Political and Social Research, University of Michigan, and from the Zentralarchiv für empirische Sozialforschung, University of Cologne, Germany. In addition, the Commission of the EC publishes a report on the highlights of each survey under the title of *Eurobarometer* reports.

3. The wording was as follows: "Here is a list of problems the people of [country] are more or less interested in. Could you please tell me for each problem whether you personally consider it a very important problem, important, of little importance, or not at all important? Protecting nature and fighting pollution . . ."

4. The Irish government was so concerned by the public's limited environmental interest relative to other Europeans that it created a new information office in the mid-1980s to stimulate greater environmental awareness in Ireland.

5. The questions read: "Sometimes measures that are designed to protect the environment cause industries to spend more money and therefore raise their prices. Which do you think is more important: protecting the environment or keeping prices down?"; and "Here are two statements which people sometimes make when discussing the environment and economic growth: (A) Protection of the environment should be given priority even at the risk of holding back economic growth;

(B) Economic growth should be given priority, even if the environment suffers to some extent. Which of these statements comes closest to your own point of view?"

6. The Gallup *Health of the Planet Survey* (1992) replicated the growth question and found the following percentages choosing the environmental option: Denmark, 77 percent; West Germany, 73 percent; Ireland, 65 percent; the Netherlands, 58 percent; and Great Britain, 56 percent.

7. The question read: "I would like to give you certain opinions which are often expressed about the problems of the environment. Which of these opinions are you most in agreement with? (1) Development of the economy should take priority over questions of the environment; (2) Sometimes it is necessary to choose between economic development or protection of the environment; and (3) Protecting the environment and preserving natural resources are essential to economic development."

8. For instance, citizen participation in campaign rallies or working for political campaigns, two voluntary political activities for which participation is regularly and strongly mobilized by political parties, is generally comparable to this voluntary involvement in environmental actions (see Barnes et al. 1979).

9. The question read: "There are a number of groups and movements seeking the support of the public. For each one of the following movements, can you tell me whether you approve (strongly or somewhat) or you disapprove (somewhat or strongly)? Nature protection associations . . . The ecology movement . . ." In most nations the questionnaire included specific examples of groups. In Britain, for instance, WWF and FoE were used as examples; in Germany it was WWF and BUND for conservation groups, and the Green Party for ecology groups. Nonetheless, the emphasis of the question was on the movement overall. The question battery also included "movements concerned with stopping the construction or use of nuclear power plants" and "anti-war and anti-nuclear weapons movements." For information on these other movements see Inglehart (1990, chap. 11).

10. The question read: "For each one of these following movements, can you tell me whether you are a member, or might join, or would certainly not join? Nature protection associations . . . The ecology movement . . ."

11. These comparisons are based on the responses of five nations included in Eurobarometer surveys in 1982 and 1989.

12. Several theorists have also suggested that religion or religiosity should be linked to support for the environment. In empirical terms, however, religion generally does not emerge as a significant predictor (Inglehart 1990, chap. 11; Shaiko 1989).

13. The Eurobarometers listed Inglehart's (1990) four-item measure. Respondents were asked the following question: "There is a lot of talk these days about what this country's goals should be for the next ten or fifteen years. On this card are listed some of the goals that different people say should be given top priority. Would you please say which one of them you consider most important in the long run? And what is next most important? (1) Maintaining order in the nation; (2) Giving the people more say in important government decisions; (3) Fighting rising prices; or (4) Protecting freedom of speech." Respondents who selected the first and third options were coded as having materialist values; those who selected the second and fourth options were coded as postmaterialists; those who chose different combinations were coded as having mixed values.

14. Our goal is not to provide an extensive empirical model of support for environmental groups. For a sophisticated attempt to provide such a fully specified model see Rohrschneider (1990).

15. The old middle class is defined as professionals and business owners; the new middle class includes white-collar salaried employees, such as office workers and government employees.

16. In some nations the social-class variable appears to have a strong impact on group involvement. However, close inspection of these cases usually finds that this class correlation arises because of distinct opinions among retirees, farmers, or other peripheral occupational groups. The middle-class/working-class dichotomy remains modest in most nations.

Chapter 4: The Organization of Environmentalism

1. In Germany, the Working Group for Environmental Questions provides an informal working group between the government and environmental groups; in Denmark, the Green Contact Committee performs a similar function.

2. We conducted a formal interview with the head of the French office and met informally with representatives of both the British and German offices. Therefore only the French organization was included in our data base of environmental organizations.

3. This figure also includes the environmental groups from the appendix who were contacted as part of this study but not formally interviewed ($N = 86$).

4. We include all mass-membership groups, federations with direct members, and elite membership groups listed in the appendix and exclude federations of groups and research institutes.

5. The list of items was drawn from Lowe and Goyder (1983) and was selected to present a mix of material, purposive, and solidarity benefits (Clark and Wilson 1961).

6. For example, membership in a few conservation groups (such as Association for the Preservation of Natural Monuments in Holland or the National Trust in Britain) provides free access to wildlife and nature preserves, and as a consequence these organizations have large memberships. The Danish Association for Nature Conservation (DN) presents an exceptionally complex case. To increase their membership base, DN mobilized a near army of elderly members scattered across the nation and offered them a capitation fee for each new member they recruited into the organization.

7. Such large membership numbers for the DNR, Friluftradet, and other federations are, however, somewhat deceiving, because they include membership in groups concerned only indirectly with environmental matters, such as alpine hiking groups, bicycling associations, hunting associations, and youth groups.

8. These figures are based on all of the groups listed in the appendix. An exact total is difficult to calculate because of imprecision in the reporting of individual group membership and the potential double-counting of members in specific groups and umbrella organizations. A membership survey by the Danish Ornithological Society, for example, found that almost half of their members also belonged to other environmental groups.

9. The Pearson correlation between the environmental membership totals in table 4.3 (adjusted by national population size) and GNP/capita is .52 ($N = 9$). Other hypothesized causal factors are correlated to national membership, but none of these relationships is statistically significant: population density ($r = .39$), the extent of neocorporatist interest group structures (.43), and social spending as a percentage of GNP (.45).

10. Although money is easily counted, there is still considerable imprecision in calculating comparable operating budgets for the groups we surveyed. One basic problem involved the complex financial structure of some environmental groups. Many European nations provide tax incentives (rebates on VAT, tax exemptions, government supplements, etc.) to nonprofit voluntary groups (charities) that are not engaged in political activities. Several activist groups have established parallel institutions to accommodate this government requirement, one to benefit from tax provisions and one to pursue political goals, both functioning de facto as a single unit. In Britain, for instance, one can contribute to the Greenpeace Environmental Trust, which is tax-exempt, or to Greenpeace Limited, which is the activist arm of the organization. Another complicating factor is that some organizations perform contract and consulting work that is supported by designated, non-reoccurring funds. Many groups administer programs that are created and funded by the government. Some organizations include these irregular sources of income in their annual income estimates, others do not. In Germany, Denmark, and Holland, an unemployed person can work for a nonprofit voluntary association and receive unemployment compensation as wages; in Italy and Belgium one can work for an environmental group as an alternative to military service. In the case of Robinwood, this indirect financial support from the government is nearly double their other income. In other instances, groups might include the expenses involved in the administration of youth branches or nature reserves, whereas other groups pursue these activities through separate organizational structures. Finally, groups differ in their fiscal years and the efficiency of their bookkeeping. Some groups had budget figures that were several years out of date; others had current figures immediately available. Our interviews probed for estimates of standard operating budgets, which were verified against published sources when possible. With some minor corrections, these are the figures reported in the table.

11. Among the forty-eight mass-membership groups included in our survey, 67 percent ranked membership dues as their primary source of income, followed by individual gifts and the sale of group materials.

12. In a 1987 survey of British environmental groups, two-thirds of the groups felt that merchandising would become an increasingly important source of their income (Elkington et al. 1988, 19).

13. Throughout this study we use .10 as the threshold for testing statistical significance. This value was chosen because these groups represent virtually the entire universe of national environmental interest groups. If one adjusts for the sample-to-universe ratio, a .10 significance level is more conservative than the conventional .05 threshold used with large sample surveys.

14. The top panel in the figure combines the number of full-time employees and part-time employees, calculated so that two part-time employees equals one full-time equivalent (FTE).

15. The nature of a group also affects its staffing patterns; umbrella organizations and research institutes rely almost exclusively on paid employees, while mass-membership groups vary in their staffing patterns.

16. The Weber-Michels imperative was also substantiated by empirical research on American voluntary associations. For example, Jeffrey Berry's study of American public interest groups found that "in a pattern similar to private interest groups, public interest lobbies are strongly oligarchic. If anything, this tendency may be even more pronounced in public interest groups, where the democratic mold structure, much less real participatory democracy, is largely absent" (1977, 193). Berry also concluded that these oligarchic tendencies appeared more pronounced in consumer and environmental groups; Kay Schlozman and John Tierney (1986) found that American public interest groups are even more oligarchic than private interest associations.

17. In some cases, the legal definition of a group determines its structure. In Holland, for example, a voluntary association can constitute itself as a foundation (*Stichting*), which limits the status of its members, or it can be formed as an association (*Vereniging*), which grants members a legal role in the operation of the organization.

18. Most of the umbrella organizations attribute greater decision-making authority to their members, but in this case the members are themselves environmental groups represented by their directors. In other words, the board of directors of these umbrella organizations is equivalent to the membership.

19. The lack of membership participation may also indicate an uncertain commitment to participatory politics among the leadership of these interest groups. One set of questions in the survey assessed the institutional norms of the representatives of mass-membership groups. From one perspective, these data display considerable support for participatory norms: 74 percent feel that the vitality of the group depends on the participation of its members, and 77 percent believe there are real opportunities for members to participate in group decision making. Yet many group representatives also endorse what might be described as elitist norms: over half (58 percent) say it is their duty to safeguard the group's interest and not just to represent their members. A substantial number think that few members have time to participate in the group (88 percent) or held strong views about what the group should be doing (55 percent). Clearly, these environmental elites do not wholly subscribe to the radical participatory rhetoric that is most closely identified with the German Green Party (see, e.g., Kitschelt 1989; Poguntke 1993). Instead, their opinions are divided between support for membership involvement and authoritative decision making.

20. RM theory implies that large groups should be centralized and that centralized groups should be large. In contrast, we find that centralized decision making is much more highly correlated with ideology ($r = .45$) than with the resources of an organization: membership ($r = .00$), full-time staff (.29), and annual budget (.00).

21. These decision-making processes are largely based on discussion and consensus building among groups rather than on formal votes. Similar patterns are found in local environmental initiatives in the United States (see Barkan 1979; Downey 1986).

22. The interviewer coded this measure of conflict based on our discussions of the internal structure and decision-making processes of the group. About half of the groups report virtually no sources of disagreement within their organization, while about one-fifth acknowledge at least occasional conflict on one or more major issues.

23. A multiple regression analysis predicting the level of internal conflict finds a strong independent impact for conservation/ecologist orientation; the relationship between structure and conflict drops to insignificance: ideology (β = .33) and structure (β = .09). The Multiple R was .38.

24. One might rightly argue that the industry analogy is more accurate than the single "bigger is better" model sometimes implied by RM theory. The retail clothing industry, for example, distributes its wares through a variety of department stores, specialty shops, discount houses, and other vendors; individuals similarly buy into the environmental movement in a variety of forms. Marketing scholars would explain the diversity of the clothing industry as a function of multidimensional consumer preferences, and a similar situation applies for public interest groups.

25. This economic analogy is useful in depicting part of the dynamic within the environmental movement, but politics (fortunately) is not the same as economics. Subsequent chapters will argue that the identity of most political groups is not simply a marketing decision. The environmental identity of a group flows from the political values of the individuals who founded the organization, the activists who presently direct it, and the supporters who have been recruited by the group. The underlying logic of political action, especially for a social movement, is to capture more support for one's political viewpoint, not just to adopt the viewpoint that can yield more resources for the organization.

Chapter 5: Environmental Elites

1. In passing we should note that ecology groups are not completely without ties to social elites. For instance, Jonathon Porritt, the former head of UK-FoE, will assume a hereditary title, and there are other sons and daughters of established elites within the movement.

2. For example, 67 percent of ecologists come from professional occupations, and 29 percent have graduate-level degrees. Ecologists are also only slightly younger (average age = 37) than the overall environmentalist sample.

3. The survey asked about length of employment and age, but not when each individual first entered the labor force. We have attempted to estimate the age of entry by using the level of schooling as a surrogate measure. Consequently, our calculations are approximations.

4. A further illustration might explain this apparent anomaly. Begin with the assumption that positions in conservation groups are twice as numerous as positions in ecology groups, for example, a 100/50 ratio. If both types of groups recruited an equal percentage from the other, for example, 10 percent, this would mean 10 conservationists came from ecology groups (10/100) and 5 ecologists came from conservation groups (5/50).

5. The generally leftward leanings of environmentalists are also confirmed in Kitschelt and Hellemans's study of green party activists (1990, chap. 4) and in Lester Milbrath's study of environmental group members (1984, 127).

6. The leftist tendency of elites is a common pattern that holds for political parties and many other political groups (e.g., Dalton 1988, chap. 10).

7. Another way of looking at these data is to say that there is a strong relationship between the environmental orientation (conservation/ecology) of a group and the personal Left/Right position of the group's leadership (tau-c = .42); a group's environmental orientation is even more strongly related to where the respondent places the group on the Left/Right scale (tau-c = .60).

8. This gap is even wider if we include responses that a group is neither Left nor Right, because representatives of conservation groups are more likely to describe the organization as lacking a clear Left/Right identity.

9. This theoretical classification of issues was validated by a factor analysis of issue opinions. Two dimensions explain most of the variation in the positions presented in table 5.2. The first has income inequality, public ownership of industry, government management of the economy, control of multinationals, and employee self-management as the highest loading variables. The second emphasizes nuclear power, postmaterial values, the storage of nuclear wastes, limits to growth, abortion rights, and defense spending.

10. This item is based on Inglehart's (1977, 1990) basic four-item question, in which respondents are asked to rank the priority of four social goals, two reflecting materialist values and two reflecting postmaterial goals.

11. It is important to remind the reader that our survey was conducted before the Chernobyl disaster. Additionally, cross-national comparisons indicate a rough correspondence between the views of environmental leaders and their government's nuclear policy: opposition to nuclear energy is generally greatest in the European states with small or nonexistent nuclear programs.

12. There is only a moderate correlation between the conservation/ecology orientation of a group and the political views of their representatives on traditional economic issues or foreign policy issues, but stronger group differences appear on New Politics issues (Pearson correlations):

Economic and Foreign Policy		*New Politics Issues*	
Control of multinationals	.13	Sexual equality	.26
Income inequality	.17	Abortion	.40
Government in economy	.01	Limits to growth	.27
Public ownership	.35	Employee rights	.41
Defense effort	.23	Nuclear waste	.29
Disarmament	.06	Nuclear energy	.41
		Postmaterialism	.53

Chapter 6: Defining the Agenda

1. This issue reached new prominence with the Chernobyl disaster, but this event occurred after the fieldwork for our study and the Eurobarometer 25 survey were completed.

2. The question read: "Now we'd like to ask a few questions about your perceptions of the environmental problems facing [your country] today. Separate from the specific concerns of your own organization, what do you think are the two or three major environmental or conservation problems that [your country] faces at the present time?"

3. Eurobarometer 25 included a differently worded question on the importance of various environmental problems. This public opinion survey also showed a high interest in air and water pollution problems, with lower levels of concern over countryside issues (European Communities 1986, 30–34).

4. For instance, one proposed remedy for damage to lakes caused by acid rain is to restore the pH balance by massive doses of antacids; ecologists would disdain such an approach and focus on the sources of air pollution.

5. The question read: "Now we'd like to discuss briefly the activities of [your group]. . . . What are the major environmental or conservation issues that have concerned your organization in the recent past?"

Chapter 7: Alliance Patterns and Environmental Networks

1. Ironically, in the latter half of the same article Offe discusses the potential coalitions that might link NSMs to established interests on either the Left or Right.

2. For information on the alliance patterns of the women's movement, see Costain and Costain (1987), Ferree (1987), and Gelb (1989). The networks of the peace movement are discussed by Rochon (1988), Mushaben (1989), and Klandermans (ed., 1991).

3. The question read: "Now we'd like to get your ideas about how well various institutions are addressing the important environmental issues facing [your nation] today. Would you indicate whether you think each of the groups I list is doing an excellent, good, fair, or poor job on these issues?" For these analyses the responses were coded: (1) poor, (2) fair, (3) good, and (4) excellent.

4. This item measures evaluations of the green parties in Belgium, France, and West Germany, as well as the following New Left parties: the Italian Radicals, Danish Radical Venstre, Dutch Pacifist Socialist Party (PSP), and Greek KKE-Interior.

5. The relevant agencies were the Ministry of Environment in France, Italy, Luxembourg, and Denmark; the Department of the Environment, Ireland and Britain; Public Health and Environment, Belgium; Ministry of Housing, Physical Planning, and Environment, Holland; Ministry of Interior, Germany; and Physical Planning and Environment, Greece.

6. Eurobarometer 25 (April 1986) included the following question: "Do you know if [in your country] the responsible authorities are concerned with the protection of the environment? (If yes), do you think the authorities are doing an effective job or not?" Responses were coded: (1) authorities concerned and effective, (2.5) concerned but not effective, and (4) not concerned. Also see Commission of the European Community (1986).

7. The House of Lords has taken an active role in focusing public attention on environmental issues and publicizing the issue throughout the past two decades (Lowe and Goyder 1983, chap. 4). In addition, many members, such as Lords Sanford, Beaumont, and Kennet, are strong advocates and participants in the conservation movement. Ironically, however, one of our British respondents claimed that the House of Lords is sympathetic to conservation issues because the majority of the peerage live on estates where the land is owned by the National Trust.

8. The relevant office in Belgium, the Netherlands, Italy, Luxembourg, Denmark, Britain, and Greece was the prime minister; in France, the president; in Germany, the chancellor; and in Ireland, the taoesich.

9. The logic held that since the appliances used less water and electricity, they made fewer demands on natural resources and created less waste water (Elkington et al. 1988).

10. The relevant associations are: France, National Council of French Industrialists (CNPF); Belgium, Belgian Business Federation; Netherlands, Union of Dutch Enterprises (VNO); Germany, Federation of German Industry (BDI); Italy, Confidustria; Denmark, Employers Association; Britain, Confederation of British Industry (CBI); and Greece, Confederation of Greek Industrialists.

11. The respective labor union federations are: France, French Democratic Confederation of Labor (CFDT); Belgium, Federation of Workers in Belgium (FGTB); Netherlands, Dutch Federation of Labor (FNV); Germany, German Labor Federation (DGB); Italy, Sindicati; Denmark, Landsorganization (LO); Britain, Trade Union Congress (TUC); Greece, Confederation of Greek Workers.

12. In each nation the questionnaire simply referred to newspapers in general. The references to television varied: France, TF1/A2/FR3; Germany, ARD and ZDF; Italy, RAI; Britain, BBC; Greece, ERT1, 2; and simply "television" in the other nations.

13. Respondents mentioned specific groups, which were then coded according to the environmental orientation of the organization.

14. Several federations, such as the Friluftradet, FFSPN, DNR, and LMO, are informal information-gathering and coordinating bodies for national and regional groups. Their membership is often large and diverse (Friluftradet has seventy-two member groups, FFSPN has more than a hundred member groups, and DNR includes ninety-five groups), ranging from ecologist groups to bicycling and hiking associations. Other federations, such as the SNM in Holland, the Working Group for Environmental Questions in Germany, the Green Alliance in Denmark, and CoEnCo in Britain, represent environmentalists before government agencies.

15. Chapter 5 shows that environmentalists hold very positive feelings toward other NSMs. In addition, other researchers find that there are strong political, personal, and resource ties binding these New Social Movements together (Mushaben 1989; Kriesi 1988; Klandermans 1990).

16. The following table provides the relationship (eta coefficient) between ecologist/conservationist orientations and nation as predictors of the performance of various political actors.

	Orientation	Nation
Business Association	.35	.20
Chief executive	.35	.48
Ministry of Industry	.32	.26
Conservative Party	.31	.37
Ministry of the Environment	.17	.52
Parliament	.14	.35

Chapter 8: Patterns of Action

1. Nearly all groups (87 percent) responded that they were "very active" or "somewhat active" in trying to influence national policy.

2. The question read: "Now speaking in more general terms, this card lists different means of action that some groups use to influence politics. For each one, would you indicate whether your group uses it often, sometimes, rarely, or never?" This battery is an adaptation of a question used by Frank Wilson (1987) in his study of French interest groups, and we would like to acknowledge his assistance and advice.

3. Although we must be cautious about comparing our French environmental groups to Frank Wilson's general study of French interest groups because of the small number of cases involved, these comparisons show that French environmental groups are more likely to pursue contacts with the media and utilize unconventional tactics; they are also less likely to engage in ministerial contacts (Wilson 1987; see also Schlozman and Tierney 1986 and Walker 1991 for evidence on citizen interest groups in the United States).

4. The questions were asked separately as part of an extended battery on each political tactic. The basic question read: "In your opinion, how effective is [tactic] as a method for [the group] to use in influencing policy making: very effective, somewhat, not very, or hardly at all."

5. The relevant agencies were the Department of the Environment in Great Britain (formed in 1970); the Ministry of Environment in France (1971); the Ministry of the Environment in Denmark (1971); the Ministry of Housing, Physical Planning, and Environment in the Netherlands (1971); the Ministry of Physical Planning and Environment in Greece (1980); the Ministry of the Environment in Luxembourg (1973); the Department of the Environment in Ireland (1972); Public Health and Environment in Belgium (1974); and the Ministry of the Environment in Italy (1986). In Germany the Ministry of the Interior was responsible for environmental policy at the time of our survey; later in 1986 a separate Ministry of the Environment and Nuclear Safety was created.

6. Jack Walker's (1991) research suggests that the creation of new citizen groups may actually be a consequence of the creation of new government programs or government agencies.

7. In France the committees are the Conseil National pour la Protection de la Nature and the Haut Comité de l'Environnement. In Belgium the government committees are the Conseil Supérieur de la Conservation de la Nature and the Conseil Supérieur pour la Protection de l'Environnement. In 1986 the Italian government created a similar advisory body, the Consiglio Nazionale dell'Ambiente.

8. This rejection of committee work must be taken with a grain of salt, however. Lowe and Goyder (1983, chap. 7) observe that FoE has been willing to engage in other committee work in the past, and it has continued to engage in such activities since our interview, albeit on a limited scale.

9. This is reflected in environmentalists' open doubts about protest activities as a method of policy influence (see fig. 8.1).

10. Several of our respondents also noted that with the accumulation of protest experiences, it becomes more difficult for groups to think up original campaigns to capture public and media interest. A banner attached to a polluting smokestack generates widespread publicity the first time but loses its impact after repeated occurrences.

11. Ministerial contact is measured by the amount of formal meetings with ministry officials, as presented in table 8.2; the commissions measure is the count of the number of governmental commissions or committees in which the group participates; the media activity variable is the amount of media contact from table 8.2; and the protest measure is an additive index of participation in the three challenging actions of table 8.2 (demonstrations, legal actions, or blocking activities).

12. See Walker (1991, chap. 6) for comparable evidence on interest groups in the United States.

13. The countries coded as neocorporatist were Belgium, the Netherlands, Germany, Luxembourg, and Denmark. Countries with leftist governments were France, Ireland, and Greece. Those with significant green/New Left parties at the time of our survey were Belgium, the Netherlands, Germany, Denmark, and Greece. System openness is a subjective judgment of the author that is intended to distinguish between systems that are willing to interact with citizen groups and those that have closed bureaucratic structures or are hostile to new interests (see, e.g., Kitschelt 1986); open systems are the Netherlands, Luxembourg, and Denmark.

14. This is not to say, however, that the same groups are active in conventional and unconventional strategies in neocorporatist systems. For instance, it may be that neocorporatist systems encourage conservation groups to become engaged in governmental lobbying while stimulating ecologist groups to be more active outside of the neocorporatist structure.

15. There is, however, a negative correlation between evaluations of the chief executive and conventional activity. We think this relationship is distorted by the restricted range of responses about executive performance; only a single group representative gave the executive either an excellent or good rating.

16. We included a large subset of variables from the above analyses. From table 8.3: size of professional staff, adequate staff size, size of membership, annual budget, the importance of membership dues, and the importance of government grants. From table 8.4: age of organization, mass-membership group, leadership control, and membership input. From table 8.5: neocorporatism, leftist government, viable New Left party, and system openness.

17. There is a strong theoretical justification for beginning each model by including environmental orientations, because we posit that the overall impact of orientations is partially direct and partially indirect, through the effect on intervening variables, such as political resources and organizational structure. Even if this forced inclusion was not used, environmental orientations would still be the first variable included in the protest model, and it would be the second variable for inclusion in the ministerial-contact and media models.

18. We computed two simple additive scales: conventional action summated the number of times a group often participated on government commissions, contacted ministerial officials, or contacted a member of parliament; and unconventional action similarly summated participation in protest activities, blocking actions, or court challenges. Among ecology groups, 35 percent engaged in all three conventional activities, and 20 percent in all three unconventional activities. Among conservation groups, the respective statistics are 77 percent and 0 percent.

19. See Przeworski and Sprague (1987) for a similar discussion of this dilemma in the context of the labor union movement.

Chapter 9: Political Parties and Environmentalism

1. For another example of this same logic see Petra Kelly (1984).

2. Lalonde was one of the founders of the French AdIT and was the environmental candidate in the 1981 presidential election. Although Lalonde's actions after the 1981 campaign left him estranged from major elements of the movement, his popular visibility as an environmentalist improved the environmental image of the government. In elections he emerged as the head of a new green party.

3. In the course of our study we spoke with a representative of SERA, who expressed open pessimism about the Labour Party's true commitment to the environment. He felt that SERA's role was to preach to the unconverted. Also see Frankland (1990) and Flynn and Lowe (1992).

4. Lowe and Goyder (1983, 128) recount how FoE in Britain felt it had been unfairly used after supporting the Labour Party in the Greater London Council elections of 1973; as a result the group claimed it would eschew further partisan ties. Our respondent recounted a similar story of FoE's negative experience in supporting some Labour candidates in 1979—with an even stronger vow to avoid direct party ties in the future.

5. More recently, the Green Party in Germany has shed some of its most unconventional organizational characteristics in an attempt to become politically more effective in the wake of their electoral loss in 1990 (Frankland and Schoonmaker 1992; Veen and Hoffmann 1992). This reinforces an earlier trend toward conventionality as the party has aged (Offe 1990).

6. At the end of the 1980s, green parties throughout Europe had won seats in the national parliament and/or the European parliament, except in Britain (because of its electoral system) and in Greece (for lack of electoral support).

7. The French Green Party has changed its name and organization several times in recent years (Prendiville and Chafer 1990). In the late 1970s several small green-oriented parties ran in national and local elections. In 1984 the largest of these groups combined to form The Greens (Les Verts). Following a surprising show of strength by The Greens in the 1989 European parliamentary elections, Brice Lalonde formed a rival green party, Génération Ecologie, that had close ties to the Socialists. In 1993 the two green parties agreed to form a common slate in the parliamentary elections, possibly a first step toward eventual merger.

8. The question read: "Are there other political parties that you feel are disinterested or even hostile to your cause?"

9. Evaluations of the environmental performance of the parties yield a similar pattern of party images (see chap. 7 and fig. 9.1).

10. Similarly, in chapter 7 we found that ecology groups are more positive in evaluating the environmental performance of green and New Left parties and more critical in judging the performance of socialist, liberal, and conservative parties. The tau-c correlation between an ecologist orientation and the environmental ratings of parties displayed in figure 7.1 are as follows: green/New Left = .20*; Socialist

Party $= -.09$; Liberal Party $= -.13$; and Conservative Party $= -.23*$. (Coefficients marked by an asterisk are significant at .10 level.)

11. There is a tau-c correlation of .30 between perceptions of the effectiveness of party work and actual contact with a party.

12. See chapter 8, and especially the discussion in connection with tables 8.4–8.6, for additional information on the construction of these measures.

13. See chapter 8, note 13.

14. There is also a slight, though statistically insignificant, tendency for ecology groups to rate party activities as more effective in influencing policy.

15. This analysis basically replicates the methodology of the analyses presented in table 8.6. The variables were listed in a regression equation, and forward selection was used to enter variables into the equation with a significance level of .10.

16. These analyses are based on overall orientations toward ecology groups and should be interpreted as such (see chap. 3 for additional discussion of this measure). In addition, the overall correlation between party preference and support for nature conservation groups is presented below (table 9.4).

17. The national averages are as follows: France (10.4 percent), Belgium (5.4), Holland (31.1), Germany (25.7), Italy (18.2), Denmark (6.8), Britain (18.2), Greece (30.0), Ireland (19.3), and Luxembourg (24.1).

18. Shortly after this survey, the PPR, the Pacifist Socialist Party, and the Communist Party of the Netherlands (CPN) federated into the "Green Left." This federation represents a clearer New Left alternative to the PvdA and was created to attract more environmentally concerned voters.

19. We also tested the apartisan hypothesis and found no clear evidence that ecologists were more likely to be nonpartisan or independents. In fact, in most nations the members or potential members of ecology groups were slightly more likely than the average voter to express a party preference.

20. These analyses use the indices of environmental group support presented in chapter 3 (see table 3.5).

21. The party patterns in figure 9.2 suggest only small differences among the Danish parties, but this results from the small number of respondents who say they are members or potential members of an ecology group. When the full range of responses is analyzed, there are sizable differences in party preferences between those who approve of the ecology movement and those who disapprove.

Chapter 10: Environmental Politics and Advanced Industrial Democracies

1. By focusing on only large national groups that deal with multiple issues, we have underplayed the diversity of environmental groups. For instance, Lowe and Goyder (1983) identify nearly eighty separate environmental groups in Britain; the German BBU once claimed nearly a thousand local groups as members; accounts of the number of local and special interest environmental groups within the European Community run into the thousands. The spokesperson for one Dutch group facetiously noted that with this continued proliferation of environmental groups, there will soon be one on each block, like McDonald's hamburger franchises.

2. The resource advantages of diversity may also apply to other public interest movements, such as consumer groups or life-style groups (see, e.g., Berry 1977; Schlozman and Tierney 1986).

3. This is the counterbalancing side to the observation that groups seeking to work with environmentalists sometimes have difficulty identifying an authoritative representative of the movement. Thus strengths and weaknesses are often represented in the same trait.

4. In addition, Gerlach (1971) observes that SMO diversity can prevent effective suppression or co-optation of the movement because there is no single group or leader to contain.

5. Only three dimensions explain at least 10 percent of the variance. The second dimension explains 13 percent of the variance and primarily reflects the size of environmental groups. The third dimension explains 11 percent of the variance and primarily taps a dimension of organizational structure.

6. The separation of structure from the goals and activities of NSMs is also increasingly apparent for the German Green Party, which is often cited as the prime example of these new political values (see Offe 1990; Frankland and Schoonmaker 1992; Veen and Hoffmann 1992).

7. The clearest example of the ability of political values to counteract a maximizing strategy comes from a subsequent research project involving Russian environmental groups. One large Russian group told our research team that it desperately needed hard currency to support its operations. When we proposed a research activity funded by an agency of the U.S. government, it rejected the offer "because we do not take money from polluters." There is more to social movements than just resource maximization.

8. Similarly, studies of individual political participation find that protest and conventional action are not mutually exclusive, and citizens who participate in one mode are actually more likely to engage in the other (Barnes et al. 1979; Kaase 1989).

References

Aberbach, Joel, and Jack Walker. 1970. Political trust and racial ideology. *American Political Science Review* 64: 1199–1219.

Agnelli Foundation. 1985. *Volontari per l'arte e per l'ambiente*. Rome: Agnelli Foundation and WWF.

Alber, Jens. 1989. Modernization, cleavage structures, and the rise of green parties and lists in Europe. In Ferdinand Müller-Rommel, ed., *New politics in Western Europe*. Boulder, Colo.: Westview Press.

Allison, Lincoln. 1975. *Environmental planning: A political and philosophical analysis*. Totowa: Rowman and Littlefield.

Anderssen, Goul. 1988. Miljopoltiske skillelinjer i Danmark. *Politica* 20: 393–413.

Andritzky, Walter, and Ulla Wahl-Terlinden. 1978. *Mitwirkung von Bürgerinitiativen und der Umweltpolitik*. Berlin: Erich Schmidt.

Bahro, Rudolf. 1984. *Building the green movement*. Philadelphia: New Society Publishers.

Baker, Susan. 1990. The evolution of the Irish ecology movement. In Wolfgang Rüdig, ed., *Green politics one*. Edinburgh: University of Edinburgh Press.

Barkan, Stephan. 1979. Strategic, tactical and organizational dilemmas of the protest movement against nuclear power. *Sociological Problems* 27: 19–37.

Barnes, Samuel, Max Kaase, et al. 1979. *Political action: Mass participation in five western democracies*. Beverly Hills: Sage.

Berger, Suzanne. 1981. *Organizing interests in Western Europe*. Cambridge: Cambridge University Press.

Berry, Jeffrey. 1977. *Lobbying for the people*. Princeton: Princeton University Press.

———. 1984. *The interest group society*. Boston: Little, Brown.

Boardman, Robert. 1981. *International organization and the conservation of nature*. Bloomington: Indiana University Press.

Bookchin, Murray. 1982. *The ecology of freedom: The emergence and dissolution of hierarchy*. Palo Alto: Cheshire.

Bramwell, Anna. 1989. *Ecology in the twentieth century: A history*. New Haven: Yale University Press.

Brand, Karl-Werner. 1982. *Neue soziale Bewegungen: Entstehung, Funktion und Perspektive neuer Protestpotentiale.* Opladen: Westdeutscher.

———. 1990. Cyclical aspects of new social movements. In Russell Dalton and Manfred Kuechler, eds., *Challenging the political order.* New York: Oxford University Press, Cambridge, Eng.: Polity Press.

———, ed. 1985. *Neue soziale Bewegungen in Westeuropa und den USA: Ein internationaler Vergleich.* New York: Campus.

Brand, Karl-Werner, Detlef Büsser, and Dieter Rucht. 1986. *Aufbruch in eine andere Gesellschaft: Neue soziale Bewegungen in der Bundesrepublik,* rev. ed. New York: Campus.

Brown, Lester. 1981. *A sustainable society.* New York: Norton.

Brown, Michael, and John May. 1989. *The Greenpeace story.* Scarborough, Ontario: Prentice-Hall Canada.

Buchholz, Rogene. 1993. *Principles of environmental management: The greening of business.* Englewood Cliffs, N.J.: Prentice Hall.

Bürklin, Wilhelm. 1984. *Grüne Politik: Ideologische Zyklen, Wähler und Parteiensysstem.* Opladen: Westdeutscher.

———. 1987a. Why study political cycles? *European Journal of Political Research* 15: 111–26.

———. 1987b. Governing Left parties frustrating the radical non-established Left. *European Sociological Review* 3: 109–26.

Caldwell, Lynton. 1984. *International environmental policy: Emergence and dimensions.* Durham: Duke University Press.

———. 1990. *Between two worlds: Science, the environmental movement, and policy choice.* Cambridge: Cambridge University Press.

Capra, Fritjof, and Charlene Spretnak. 1984. *Green politics: The global promise.* New York: Dutton.

Carson, Rachel. 1962. *The silent spring.* Boston: Houghton Mifflin.

Chafer, Tony. 1982. The anti-nuclear movement and the rise of political ecology. In Philip Cerny, ed., *Social movements and protest in France.* London: Pinter.

Chong, Dennis. 1991. *Collective action and the civil rights movement.* Chicago: University of Chicago Press.

Clark, P., and James Wilson. 1961. Incentive Systems. *Administrative Science Quarterly* 6: 129–66.

Commission of the European Communities. 1982. *The Europeans and their environment.* Brussels: Commission of the European Community.

———. 1986. *The Europeans and their environment in 1986.* Brussels: Commission of the European Community.

———. 1990. *Environmental policy in the European Community,* 4th ed. Luxembourg: Office of Official Publications on the European Communities.

Conover, Pamela, and Stanley Feldman. 1984. How people organize the political world. *American Journal of Political Science* 28: 95–126.

Conradt, David, and Russell Dalton. 1988. The West German electorate and the party system. *Review of Politics* 50: 3–29.

Conwentz, Hugo. 1909. *The care of natural monuments.* Cambridge: Cambridge University Press.

Costain, Anne, and W. Douglas Costain. 1987. Strategy and tactics of the women's movement in the United States. In Mary Katzenstein and Carol Mueller, eds., *The women's movement of the United States and Western Europe.* Philadelphia: Temple University Press.

Cotgrove, Stephen. 1982. *Catastrophe or cornucopia: The environment, politics and the future.* New York: Wiley.

Cotgrove, Stephen, and Andrew Duff. 1980. Environmentalism, middle-class radicalism, and politics. *Sociological Review* 28: 333–51.

———. 1981. Environmental values and social change. *British Journal of Sociology* 32: 92–110.

Curtis, Russell, and Louis Zurcher. 1973. Stable resources of protest movements. *Social Forces* 52: 53–61.

Dalton, Russell. 1984. Environmentalism and value change in Western democracies. Paper presented at the annual meeting of the American Political Science Association, New York.

———. 1988. *Citizen politics in Western democracies: Public opinion and political parties in the United States, Great Britain, West Germany and France.* Chatham, N.J.: Chatham House.

Dalton, Russell, Scott Flanagan, and Paul Beck, eds. 1984. *Electoral change in advanced industrial democracies: Realignment or dealignment?* Princeton: Princeton University Press.

Dalton, Russell, and Manfred Kuechler, eds. 1990. *Challenging the political order: New social and political movements in Western democracies.* New York: Oxford University Press, Cambridge, Eng.: Polity Press.

DeClair, Edward. 1986. Nonparticipation in the French environmental movement. Report of the Environmental Movements in Western Democracies Project, Tallahassee, Fla. Photocopy.

della Porta, Donatella, and Dieter Rucht. 1994. Left-libertarian movements in context. In J. Craig Jenkins and Bert Klandermans, eds., *The politics of social protest.* Minneapolis: University of Minnesota Press.

Devall, Bill. 1988. *Simple in means, rich in ends: Practicing deep ecology.* Salt Lake City, Utah: Peregrine Smith.

Diamond, Irene, and Gloria Orenstein. 1990. *Reweaving the world: The emergence of ecofeminism.* San Francisco: Sierra Club.

Diani, Mario. 1988. *Isole nell'Archipellago: Il movimento ecologista in Italia.* Bologna: Societa Editrice il Mulino.

———. 1990a. The Italian ecology movement: From radicalism to moderation. In Wolfgang Rüdig, ed., *Green politics one.* Edinburgh: University of Edinburgh Press.

———. 1990b. The network structure of the Italian ecology movement. *Social Science Information* 29: 5–31.

———1994. *The green connection: Structures of environmental action in Italy.* Edinburgh: University of Edinburgh Press.

Diani, Mario, and Giovanni Lodi. 1988. Three in one: Currents in the Milan ecology movement. In Bert Klandermans, Hanspeter Kriesi, and Sidney Tarrow, eds., *From structure to action.* Greenwich, Conn.: JAI Press.

Dominick, Raymond. 1992. *The environmental movement in Germany: Prophets and pioneers, 1871–1971.* Bloomington: Indiana University Press.

Donati, Paolo. 1984. Organization between movement and institution. *Social Science Information* 23: 837–59.

Doughty, Robin. 1975. *Feather fashions and bird protection: A study in nature conservation.* Berkeley: University of California Press.

Downey, Gary. 1986. Ideology and the clamshell identity. *Social Problems* 33: 101–17.

Downing, Paul, and Gordon Brady. 1974. The role of citizen interest groups in environmental policy formation. In M. White, ed., *Nonprofit firms in a three-sector economy*. Washington: Urban Institute.

Duclos, D., and J. Smadja. 1985. Culture and the environment in France. *Environmental Management* 9: 135–40.

Dunlap, Riley. 1992. Trends in public opinion toward environmental issues, 1965–1990. In Riley Dunlap and Angela Mertig, eds., *American environmentalism: The U.S. environmental movement, 1970–1990*. Philadelphia: Taylor and Francis.

Dunlap, Riley, and Angela Mertig, eds. 1992. *American environmentalism: The U.S. environmental movement, 1970–1990*. Philadelphia: Taylor and Francis.

Dunlap, Riley, and K. van Liere. 1977. The new environmental paradigm. *Journal of Environmental Education* 9: 10–19.

Duyvendak, Jan, and Rob van Huizen. 1982. *Nieuwe Sociale Bewegingen in Nederland*. Zwolle: Stichting Voorlichting aktievegeweldloosheid.

Eckersley, Robyn. 1992. *Environmentalism and political theory: Toward an ecocentric approach*. Albany: State University of New York Press.

Elkington, John, Tom Burke, and Julia Hailes, eds. 1988. *Green pages*. London: Routledge.

Ewringmann, Dieter, and Klaus Zimmermann. 1978. Umweltpolitische Interessenanalyse der Unternehmen, Gewerkschaften und Gemeinden. In Martin Jänicke, ed., *Umweltpolitik: Beiträge zur Politologie des Umweltschutzes*. Opladen: Westdeutscher.

Eyerman, Ron, and Andrew Jamison. 1989. Environmental knowledge as an organizational weapon: The case of Greenpeace. *Social Science Information* 28: 99–120.

Ferree, Myra Marx. 1987. Feminist politics in the U.S. and West Germany. In Mary Katzenstein and Carol Mueller, eds., *The women's movement of the United States and Western Europe*. Philadelphia: Temple University Press.

Ferree, Myra Marx, and Bess Hess. 1985. *Controversy and coalition: The new feminist movement*. Boston: Twayne.

Ferree, Myra Marx, and F. Miller. 1985. Mobilization and meaning. *Sociological Inquiry* 55: 38–61.

Fireman, Bruce, and William Gamson. 1979. Utilitarian logic in the resource mobilization perspective. In Mayer Zald and John McCarthy, eds., *The dynamics of social movements*. Cambridge, Mass.: Winthrop.

Fitter, Richard, and Sir Peter Scott. 1978. *The penitent butchers*. London: The Fauna and Flora Preservation Society.

Flacks, Richard. 1967. The liberated generation. *Journal of Social Issues* 23: 52–75.

Flynn, Andrew, and Philip Lowe. 1992. The greening of the Tories. In Wolfgang Rüdig, ed., *Green politics two*. Edinburgh: University of Edinburgh Press.

Fogt, Helmut. 1987. Die Grünen und die Neue Linke. In M. Langner, ed., *Die Grünen auf dem Prüfstand*. Bergisch-Gladbach: Luebbe.

Foratom. 1979. *Acceptance of nuclear power*. Essen: Vulkan.

Forschungsgruppe Wahlen. 1990. *Bundestagswahl 1990*. Mannheim: Forschungsgruppe Wahlen.

Foss, Daniel, and Ralph Larkin. 1986. *Beyond revolution: A new theory of social movements*. South Hadley, Mass.: Bergin and Garvey.

Frankland, Gene. 1990. Does green politics have a future in Britain? In Wolfgang Rüdig, ed., *Green politics one*. Edinburgh: University of Edinburgh Press.

Frankland, Gene, and Donald Schoonmaker. 1992. *Between protest and power: The Green Party in Germany*. Boulder, Colo.: Westview Press.

Franklin, Mark, et al. 1992. *Electoral change: Responses to evolving social and attitudinal structures in Western countries*. New York: Cambridge University Press.

Freeman, Jo. 1975. *The politics of women's liberation*. New York: Longman.

————. 1979. Resource mobilization and strategy. In Mayer Zald and John Mc-Carthy, eds., *The dynamics of social movements*. Cambridge, Mass.: Winthrop.

————. 1983. A model for analyzing the strategic options of social movement organizations. In Jo Freeman, ed., *Social movements of the sixties and seventies*. New York: Longman.

Frohlich, Norman, Joseph Oppenheimer, and Oran Young. 1971. *Political leadership and collective goods*. Princeton: Princeton University Press.

Fuchs, Dieter, and Hans-Dieter Klingemann. 1990. The Left/Right schema. In M. Kent Jennings and Jan van Deth, eds., *Continuities in political action*. Berlin: deGruyter.

Fuchs, Dieter, and Dieter Rucht. 1990. Support for new social movements in five Western European countries. Report from the Wissenschaftszentrum, Berlin. Photocopy.

Gallup Institute. 1992. *Health of the planet survey*. Princeton: George Gallup Institute.

Gamson, William. 1975. *The strategy of social protest*. Homewood, Ill.: Dorsey Press.

————. 1992. *Talking politics*. New York: Cambridge University Press.

Gamson, William, and Gadi Wolfsfeld. 1993. Movements and media as interacting systems. *Annals of the American Academy of Political and Social Sciences* 528: 114–25.

Gelb, Joyce. 1989. *Feminism and politics*. Berkeley: University of California Press.

Gerlach, Luther. 1971. Movements of revolutionary change. *American Behavioral Scientist* 14: 812–36.

Gerlach, Luther, and Virginia Hine. 1970. *People, power and change*. Indianapolis: Bobbs-Merrill.

Gorz, Andre. 1980. *Ecology as politics*. Boston: Southend Press.

Grant, Wyn, ed. 1985. *The political economy of corporatism*. London: Macmillan.

Guggenberger, Bernd, and Udo Kempf, eds. 1984. *Bürgerinitiativen und repräsentatives System*, 2d ed. Opladen: Westdeutscher.

Gundelach, Peter. 1984. Social transformation and new forms of voluntary associations. *Social Science Information* 23: 1049–81.

Gurr, T. R. 1970. *Why men rebel*. Princeton: Princeton University Press.

Haas, Peter, Robert Keohane, and Marc Levy, eds. 1993. *Institutions for the earth: Sources of effective international environmmental protection*. Cambridge: MIT Press.

Hager, Carol. 1993. Democratizing technology: Citizen movements in Germany. *Annals of the Academy of Political and Social Sciences* 528: 42–55.

Hatch, Michael. 1986. *Politics and nuclear power: Energy policy in Western Europe*. Lexington: University of Kentucky Press.

Hayden, Sherman. 1942. *The international protection of wildlife*. New York: Columbia University Press.

Hays, Samuel. 1959. *Conservation and the gospel of efficiency: The progressive conservation movement, 1890–1920*. Cambridge: Harvard University Press.

Hays, Samuel, in collaboration with Barbara Hays. 1987. *Beauty, health and permanence: Environmental politics in the United States, 1955–1985.* New York: Cambridge University Press.

Heberle, Rudolf. 1951. *Social movements: An introduction to political sociology.* New York: Appleton Century.

Hellemans, Staf. 1985. Maatschappelijke verandering op nieuwe wegen. *Socialistische Standpunten* 32: 1–16.

Hirschman, Albert. 1970. *Exit, voice, and loyalty: Responses to the decline in firms, organizations and states.* Cambridge: Harvard University Press.

Hoffer, Eric. 1951. *The true believer.* New York: Harper and Row.

Hofrichter, Jürgen, and Karlheinz Reif. 1990. Evolution of environmental attitudes in the European Community. *Scandinavian Political Studies* 13: 119–46.

Hunter, Robert. 1979. *Warriors of the rainbow: A chronicle of the Greenpeace movement.* New York: Holt, Rinehart and Winston.

Huntington, Samuel. 1981. *American politics: The promise of disharmony.* Cambridge: Harvard University Press.

Inglehart, Ronald. 1977. *The silent revolution: Changing values and political styles among Western publics.* Princeton: Princeton University Press.

———. 1984. The changing structure of political cleavages in Western society. In R. Dalton, S. Flanagan, and P. Beck, eds. *Electoral change in advanced industrial democracies.* Princeton: Princeton University Press.

———. 1990. *Culture shift in advanced industrial society.* Princeton: Princeton University Press.

Inglehart, Ronald, et al. 1980. Broader power for the European Parliament? *European Journal of Political Research* 8: 113–32.

Ingram, Helen, and Dean Mann. 1989. Interest groups and environmental policy. In James Lester, ed., *Environmental politics and policy.* Durham: Duke University Press.

Irvine, Sandy, and Alec Ponton. 1988. *A green manifesto.* London: Optima.

Jamison, Andrew, Ron Eyerman, and Jacqueline Cramer. 1990. *The making of the new environmental consciousness: A comparative study of environmental movements in Sweden, Denmark and the Netherlands.* Edinburgh: University of Edinburgh Press.

Jenkins, J. Craig. 1983. Resource mobilization theory and the study of social movements. *Annual Review of Sociology* 9: 527–53.

Jenkins, J. Craig, and Bert Klandermans, eds. 1994. *The politics of social protest: Comparative studies of states and social movements.* Minneapolis: University of Minnesota Press.

Judge, David, ed. 1993. *A green dimension for the European Community.* London: Frank Cass.

Kaase, Max. 1989. Mass participation. In M. Kent Jennings and Jan van Deth, eds., *Continuities in political action.* Berlin: deGruyter.

———. 1990. Social movements and political innovation. In Russell Dalton and Manfred Kuechler, eds., *Challenging the political order.* New York: Oxford University Press, Cambridge, Eng.: Polity Press.

Kamieniecki, Sheldon, ed. 1993. *Environmental politics in the international arena: Movements, parties, organizations and policy.* Albany: SUNY Press.

Katzenstein, Mary, and Carol Mueller, eds. 1987. *The women's movements of the United States and Western Europe: Consciousness, political opportunity and public policy.* Philadelphia: Temple University Press.

Kazis, Richard, and Richard Grossman. 1982. *Fear at work: Job blackmail, labor and the environment*. New York: Pilgrim Press.

Kelly, Petra. 1984. *Fighting for hope*. Boston: South End Press.

Keniston, Kenneth. 1968. *Young radicals*. New York: Harcourt, Brace and World.

Kessel, Hans, and Wolfgang Tischler. 1982. *International environmental survey: Umweltbewußtsein im internationalen Vergleich*. Berlin: Wissenschaftszentrum Berlin.

Kielbowicz, Richard, and Clifford Scherer. 1986. The role of the press in the dynamics of social movements. *Research in social movements: Conflict and change*. Greenwich, Conn.: JAI Press, 71–96.

Kitschelt, Herbert. 1986. Political opportunity structures and political protest. *British Journal of Political Science* 16: 58–95.

———. 1988. Left-libertarian parties: Explaining innovation in competitive party systems. *World Politics* 40: 194–234.

———. 1989. *The logics of party formation: Ecological politics in Belgium and West Germany*. Ithaca: Cornell University Press.

———. 1990. La gauche libertaire et les écologistes français. *Revue Française de Science Politique* 40: 339–65.

Kitschelt, Herbert, and Staf Hellemans. 1990. *Beyond the European Left: Ideology and political action in the Belgian ecology parties*. Durham: Duke University Press.

Klandermans, Bert. 1986. New social movements and resource mobilization. *International Journal of Mass Emergencies and Disasters* 4: 13–39.

———. 1988. The formation and mobilization of consensus. In Bert Klandermans, Hanspeter Kriesi, and Sidney Tarrow, eds., *From structure to action*. Greenwich, Conn.: JAI Press.

———. 1990. Linking the "old" and "new": Movement networks in the Netherlands. In Russell Dalton and Manfred Kuechler, eds., *Challenging the political order*. New York: Oxford University Press, Cambridge, Eng.: Polity Press.

———. 1991. The peace movement and social movement theory. In Bert Klandermans, ed., *Peace movements in Western Europe and the United States*. Greenwich, Conn.: JAI Press.

———, ed. 1989. *Organizing for change*. A Research Annual in the International Social Movement Research Series, vol. 2. Greenwich, Conn.: JAI Press.

———, ed. 1991. *Peace movements in Western Europe and the United States*. A Research Annual in the International Social Movement Research Series, vol. 3. Greenwich, Conn.: JAI Press.

Klandermans, Bert, Hanspeter Kriesi, and Sidney Tarrow, eds. 1988. *From structure to action: Comparing social movement research across cultures*. A Research Annual in the International Social Movement Research Series, vol. 1. Greenwich, Conn.: JAI Press.

Klingemann, Hans-Dieter. 1979. Measuring ideological conceptualizations. In Samuel Barnes, Max Kaase, et al., *Political action*. Beverly Hills: Sage.

———. 1985. Umweltproblematik in den Wahlprogrammen der etablierten politischen Parteien in der Bundesrepublik Deutschland. In Rudolf Wildenmann, ed., *Umwelt, Wirtschaft, Gesellschaft*. Stuttgart: Staatsministeriums Baden-Württemberg.

Kornhauser, William. 1959. *The politics of mass society*. New York: Free Press.

Kriesi, Hanspeter. 1988. Local mobilization for the people's petition of the Dutch peace movement. In Bert Klandermans, Hanspeter Kriesi, and Sidney Tarrow, eds., *From structure to action*. Greenwich, Conn.: JAI Press.

————. 1994. The political opportunity structure of new social movements. In J. Craig Jenkins and Bert Klandermans, eds., *The politics of social protest*. Minneapolis: University of Minnesota Press.

————, ed. 1985. *Bewegungen in der Schweizer Politik*. Frankfurt: Campus.

Kriesi, Hanspeter, and Philip van Praag. 1987. Old and new politics. *European Journal of Political Research* 15: 319–46.

Kuechler, Manfred. 1984. Die Friedensbewegung in der BRD. In J. Falter, C. Fenner, and M. Greven, eds., *Politische Willenbildung und Interessenvermittlung*. Opladen: Westdeutscher.

————. 1991. Public perceptions of parties' economic competence. In Helmut Norpoth, Michael Lewis-Beck, and Jean-Dominique Lafay, eds., *Economics and politics*. Ann Arbor: University of Michigan Press.

Langguth, Gerd. 1986. *The green factor in German politics: From protest movement to political party*. Boulder, Colo.: Westview Press.

Leggewie, Claus. 1985. Propheten ohne Macht. In Karl-Werner Brand, ed., *Neue soziale Bewegungen in Westeuropa und den USA*. Frankfurt: Campus.

Lehmbruch, Gerhard, and Philippe Schmitter, eds. 1982. *Patterns of neo-corporatist policy making*. Beverly Hills: Sage.

Levy, Marc. 1991. The greening of the United Kingdom: An assessment of competing explanations. Paper presented at the annual meeting of the American Political Science Association, Washington, D.C.

Linse, Ulrich. 1983. *Zurück o Mensche zur Mutter Erde: Landkommunen in Deutschland 1890–1933*. Munich: DTV.

Lipset, Seymour. 1981. The revolt against modernity. In Per Torsvik, ed., *Mobilization, center-periphery structures and nation-building*. Bergen: Universitetsforlaget.

Lipsky, Michael. 1968. Protest as a political resource. *American Political Science Review* 62: 1144–58.

Lowe, Philip. 1983. Values and institutions in the history of British nature conservation. In A. Warren and F. Goldsmith, eds., *Conservation in perspective*. New York: Wiley.

Lowe, Philip, and Jane Goyder. 1983. *Environmental groups in politics*. London: Allen and Unwin.

McAdam, Doug. 1982. *Political process and the development of black insurgency, 1930–1970*. Chicago: University of Chicago Press.

————. 1983. The decline of the civil rights movement. In Jo Freeman, ed., *Social movements of the sixties and seventies*. New York: Longman.

McAdam, Doug, et al. 1988. Social movements. In Neil Smelser, ed., *Handbook of sociology*. Newbury Park, Calif.: Sage.

McAdam, Doug, and Dieter Rucht. 1993. The international diffusion of social protest. *Annals of the Academy of Political and Social Sciences* 528: 56–74.

McCarthy, John, and Mayer Zald. 1973. *The trends of social movements in America*. Morristown, Penn.: General Learning Press.

————. 1977. Resource mobilization and social movements. *American Journal of Sociology* 82: 1212–41.

McClosky, Herbert, Paul Hoffman, and Rosemary O'Hara. 1960. Issue conflict and consensus among party leaders. *American Political Science Review* 54: 406–27.

McCormick, John. 1991. *British politics and the environment*. London: Earthscan.

————. 1992. *The global environmental movement*. London: Belhaven Press.

Manes, Christopher. 1990. *Green rage: Radical environmentalism and the unmaking of civilization*. Boston: Little, Brown.

Marsh, Jan. 1982. *Back to the land: The pastoral impulse in Victorian England.* New York: Quartet.

Meadows, Donella, et al. *The limits to growth.* Washington, D.C.: Potomac Associates.

Melucci, Alberto. 1980. The new social movements. *Social Science Information* 19: 199–226.

———. 1984. An end to social movements. *Social Science Information* 23: 819–35.

———. 1989. *Nomads of the present.* London: Hutchinson Radius.

Merchant, Carolyn. 1980. *The death of nature: Women, ecology and the scientific revolution.* San Francisco: Harper and Row.

Michels, Roberto. 1954. *Political parties.* Glencoe: Free Press.

Milbrath, Lester. 1984. *Environmentalists: Vanguard for a new society.* Albany: SUNY Press.

———. 1989. *Envisioning a sustainable society.* Albany: SUNY Press.

Mirowski, John, and Catherine Ross. 1981. Protest group success. *Sociological Focus* 14: 177–92.

Mitchell, Robert. 1979. National environmental lobbies and the apparent illogic of collective action. In C. Russell, ed., *Collective decision making.* Baltimore: Johns Hopkins University Press.

Moe, Terry. 1980. *The organization of interests.* Chicago: University of Chicago Press.

Moltke, Konrad von, and Nico Visser. 1982. *Die Rolle der Umweltschutzverbände im politischen Entscheidungsprozess der Niederlande.* Berlin: Erich Schmidt.

Morris, A. 1984. *The origins of the civil rights movement.* New York: Free Press.

Morris, A., and Carol Mueller. 1982. *Frontiers in social movement theory.* New Haven: Yale University Press.

Müller-Rommel, Ferdinand. 1985. New social movements and smaller parties. *West European Politics* 8: 41–54.

———. 1990. New political movements and "New Politics" parties in Western Europe. In Russell Dalton and Manfred Kuechler, eds., *Challenging the political order.* New York: Oxford University Press.

———. 1993. *Grüne Politik.* Opladen: Westdeutscher.

———, ed. 1989. *New politics in Western Europe: The rise and success of green parties and alternative lists.* Boulder, Colo.: Westview Press.

Murphy, Detlef, et al. 1981. Haben "links" und "rechts" noch Zunkunft? *Politische Vierteljahresschrift* 22: 398–414.

Mushaben, Joyce. 1989. The struggle within. In Bert Klandermans, ed. *Organizing for change.* Greenwich, Conn.: JAI Press.

Naess, Arne. 1988. Deep ecology and ultimate premises. *Ecologist* 18: 128–31.

Neidhardt, Friedhelm. 1985. Einige Ideen zu einer allgemeinen Theorie sozialer Bewegungen. In Stefan Hradil, ed., *Sozialstruktur im Umbruch.* Opladen: Westdeutscher.

Nelkin, Dorothy, and Michael Pollak. 1981. *The atom beseiged.* Cambridge: MIT Press.

Nicholson, Max. 1970. *The environmental revolution: A guide for the new masters of the world.* New York: McGraw-Hill.

———. 1987. *The new environmental age.* Cambridge: Cambridge University Press.

Nisbit, Robert. 1982. *Prejudices: A philosophical dictionary.* Cambridge: Harvard University Press.

Oberschall, A. 1973. *Social conflict and social movements*. Englewood Cliffs, N.J.: Prentice-Hall.

Offe, Claus. 1984. *Cultural contradictions of capitalism*. Cambridge: MIT Press.

———. 1985. New social movements. *Social Research* 52: 817–68.

———. 1990. Reflections on the institutional self-transformation of movement politics. In Russell Dalton and Manfred Kuechler, eds. *Challenging the political order*. New York: Oxford University Press, Cambridge, Eng.: Polity Press.

Olson, Mancur. 1965. *The logic of collective action*. Cambridge: Harvard University Press.

Paehlke, Robert. 1989. *Environmentalism and the future of progressive politics*. New Haven: Yale University Press.

Parkin, Frank. 1972. *Middle class radicalism: The social bases of the British campaign for nuclear disarmament*. Manchester, Eng.: Manchester University Press.

Parkin, Sarah. 1989. *Green parties: An international guide*. London: Heretic.

Patterson, William. 1989. Environmental politics. In Gordon Smith, William Patterson, and Peter Merkl, eds., *Development in West German politics*. London: Macmillan.

Pedersen, Mogens. 1989. The birth, the life and the death of small parties in Danish politics. In Ferdinand Müller-Rommel and Geoffrey Pridham, eds., *Small parties in Western Europe*. London: Sage.

Pepper, David. 1984. *The roots of modern environmentalism*. London: Croom Helm.

Perrow, Charles. 1979. The sixties observed. In Mayer Zald and John McCarthy, eds. *Dynamics of social movements*. Cambridge, Mass.: Winthrop.

Petulla, Joseph. 1988. *American environmental history*. Columbus, Ohio: Bobbs-Merrill.

Pierce, John, et al. 1992. *Citizens, political communication, and interest groups: Environmental organizations in Canada and the United States*. Westport, Conn.: Greenwood Press.

Piven, Frances, and Richard Cloward. 1979. *Poor people's movements: Why they succeed, how they fail*. New York: Random House.

Poguntke, Thomas. 1987. New politics and party systems. *West European Politics* 10: 76–88.

———. 1993. *Alternative politics: The German Green Party*. Edinburgh: University of Edinburgh Press.

Porritt, Jonathan. 1984. *Seeing green: The politics of ecology explained*. London: Basil Blackwell.

Porter, Gareth, and Janet Welsh Brown. 1991. *Global environmental politics*. Boulder, Colo.: Westview Press.

Prendiville, Brendan, and Tony Chafer. 1990. Activists and ideas in the green movement in France. In Wolfgang Rüdig, ed., *Green politics one*. Edinburgh: University of Edinburgh Press.

Przeworski, Adam, and John Sprague. 1987. *Paper stones: A history of electoral socialism*. Chicago: University of Chicago Press.

Putnam, Robert. 1976. *The comparative study of political elites*. Englewood Cliffs, N.J.: Prentice Hall.

Rammstedt, Otthein. 1978. *Soziale Bewegung*. Frankfurt: Suhrkamp.

Raschke, Joachim. 1985. *Soziale Bewegungen: Ein historischsystematischer Grundriß*. Frankfurt: Campus.

Rochon, Thomas. 1988. *Mobilizing for peace: The antinuclear movements in Western Europe*. Princeton: Princeton University Press.

————. 1990. Political movements and state authority in liberal democracies. *World Politics* 42: 289–99.

Rohrschneider, Robert. 1988. Citizens' attitudes toward environmental issues. *Comparative Political Studies* 21: 346–67.

————. 1989. *The greening of party politics in Western Europe*. Ph.D. diss., Florida State University.

————. 1990. The roots of public opinion toward new social movements. *American Journal of Political Science* 34: 1–30.

————. 1993a. Environmental belief systems in Western Europe: A hierarchical model of constraint. *Social Science Quarterly* 72: 3–29.

————. 1993b. New party versus Old Left realignment. *Journal of Politics* 55: 682–701.

————. 1993c. Addressing the challenge: The response of West European party systems to new social movements. *Annals of the American Academy of Political and Social Sciences* 528: 157–70.

Roth, Roland, and Dieter Rucht, eds. 1987. *Neue soziale Bewegungen in der Bundesrepublik Deutschland*. Frankfurt: Campus.

Rucht, Dieter. 1980. *Von Wyhl nach Gorleben*. Munich: Beck.

————. 1982. Neue soziale Bewegungen: Die Grenzen bürokratischer Modernisierung. In J. Hesse, ed., *Politikwissenschaft und Verwaltungswissenschaft in der Bundesrepublik*. Sonderheft, *Politische Vierteljahresschrift*. Opladen: Westdeutscher.

————. 1989. Environmental movement organizations in West Germany and France. In Bert Klandermans, ed., *Organizing for change*. Greenwich, Conn.: JAI Press.

————. 1990a. The strategies and action repertoires of new movements. In Russell Dalton and Manfred Kuechler, eds., *Challenging the political order*. New York: Oxford University Press, Cambridge, Eng.: Polity Press.

————. 1990b. Campaigns, skirmishes and battles. *Industrial Crisis Quarterly* 4: 193–222.

————, ed. 1992. *Research on social movements: The state of the art in Western Europe and the USA*. Frankfurt: Campus.

Rüdig, Wolfgang. 1991. *Anti-nuclear movements: A world survey of opposition to nuclear energy*. London: Longmans.

————, ed. 1990. *Green politics one*. Edinburgh: University of Edinburgh Press.

————, ed. 1992. *Green politics two*. Edinburgh: University of Edinburgh Press.

Rüdig, Wolfgang, and Mark Franklin. 1992. Green prospects: The future of green parties in Britain, France and Germany. In Wolfgang Rüdig, ed., *Green politics two*. Edinburgh: University of Edinburgh Press.

Rüdig, Wolfgang, and Philip Lowe. 1986. The withered greening of British politics. *Political Studies* 34: 262–84.

Ryle, Martin. 1988. *Ecology and socialism*. London: Radius.

Salisbury, Robert. 1969. An exchange theory of interest groups. *Midwest Journal of Political Science* 13: 1–32.

Scheffer, Victor. 1991. *The shaping of environmentalism in America*. Seattle: University of Washington Press.

Schlozman, Kay, and John Tierney. 1986. *Organized interests and American democracy*. New York: Harper and Row.

Schmitt, Rüdiger. 1990. *Die Friedensbewegung in der Bundesrepublik Deutschland*. Opladen: Westdeutscher.

Schmitt-Beck, Rüdiger. 1990. Über die Bedeutung der Massenmedien für soziale Bewegungen. *Kölner Zeitschrift für Soziologie und Sozialpsychologie* 42: 642–62.

———. 1992. A myth institutionalized: Theory and research on new social movements in the Federal Republic of Germany. *European Journal of Political Research* 21: 357–83.

Schnaiberg, Allan. 1980. *The environment from surplus to scarcity.* New York: Oxford University Press.

Schoenfeld, Clay, et al. 1979. Constructing a social problem. *Social Problems* 27: 38–61.

Schumacher. E.F. 1973. *Small is beautiful.* New York: Harper and Row.

Scott, Alan. 1990. *Ideology and the new social movements.* London: Unwin Hyman.

Shaiko, Ronald. 1989. The public interest dilemma. Ph.D. diss., Syracuse University.

———. 1993. Greenpeace USA. *Annals of the American Academy of Political and Social Sciences* 528: 88–100.

Sheail, John. 1976. *Nature in trust: The history of nature conservation in Britain.* London: Blackie.

Sieferle, Rolf. 1984. *Forschrittsfeinde?: Opposition gegen Technik und Industrie von Romantik bis zur Gegenwart.* Munich: Beck.

Siegmann, Heinrich. 1985. *The conflicts between labor and environmentalism in the Federal Republic of Germany and the United States.* Aldershot, Eng.: Gower.

Smelser, Neil. 1963. *Theory of collective behavior.* New York: Free Press.

Smith, Charlie-Pye, and Chris Rose. 1984. *Crisis and conservation: Crisis in the British countryside.* London: Penguin.

Snow, David, et al. 1986. Frame alignment processes, micromobilization and movement participation. *American Sociological Review* 51: 464–81.

Snow, David, and Robert Benford. 1988. Ideology, frame resonance, and participant mobilization. In Bert Klandermans et al., *From structure to action.* Greenwich, Conn.: JAI Press.

Soros, Marvin. 1994. From Stockholm to Rio: The evolution of global environmental governance. In Norman Vig and Michael Kraft, eds., *Environmental policy in the 1990s.* Washington, D.C.: Congressional Quarterly Press.

Stamp, Sir Dudley. 1969. *Nature conservation in Britain.* London: Collins.

Steedly, H. R., and J. Foley. 1979. The success of protest groups. *Social Science Research* 8: 1–15.

Steger, Mary Ann, and Stephanie Witt. 1989. Gender differences in environmental orientations. *Western Political Quarterly* 42: 627–49.

Stern, Paul, Thomas Dietz, and Linda Kalof. 1993. Value orientations, gender, and environmental concern. *Environment and Behavior* 25: 322–48.

Stevis, Dimitris. 1993. Political ecology in the semi-periphery: Lessons from Greece. *International Journal of Urban and Regional Movements* 17: 85–97.

Strodthoff, G. C., et al. 1985. Media roles in a social movement. *Journal of Communication* 35: 134–53.

Tarrow, Sidney. 1983. *Struggling to reform.* Western Societies Paper, no. 15. Ithaca: Cornell University Press.

———. 1989. *Democracy and disorder.* New York: Oxford University Press.

Tellegen, Egbert. 1981. The environmental movement in the Netherlands. In T. O'Riordan and R. Turner, eds., *Progress in resource management and environmental planning*. New York: Wiley.

————. 1983. *Milieubeweging*. Utrecht: Aula.

Thomas, Keith. 1983. *Man and the natural world*. London: Allen Lane.

Tilly, Charles. 1969. Collective violence in European perspective. In Hugh Graham and Ted Gurr, eds., *Violence in America*. New York: Bantam.

————. 1975. Revolutions and collective violence. In Fred Greenstein and Nelson Polsby, eds., *Handbook of political science,* vol. 3. Reading, Mass.: Addison-Wesley.

————. 1978. *From mobilization to revolution*. Reading, Mass.: Addison-Wesley.

Tilly, Charles, Louise Tilly, and R. Tilly. 1975. *The rebellious century*. Cambridge: Harvard University Press.

Touraine, Alain. 1981. *The voice and the eye: An analysis of social movements*. Cambridge: Cambridge University Press.

————. 1983. *Anti-nuclear protest*. Cambridge: Cambridge University Press.

Tucker, William. 1982a. The environmental era. *Public Opinion* 5: 41–47.

————. 1982b. *Progress and privilege: America in the age of environmentalism*. Garden City, N.J.: Doubleday.

Tullock, Gordon. 1971. The paradox of revolution. *Public Choice* 11:425–47.

Turk, H., and L. Zucker. 1984. Majority and organization opposition. In R. Ratcliff, ed., *Research in social movements*. Greenwich, Conn.: JAI Press.

Turner, Ralph. 1970. Determinants of social movement strategies. In T. Shibutani, ed., *Collective behavior*. Englewood Cliffs, N.J.: Prentice-Hall.

Vadrot, C. 1978. *L'écologie: Histoire d'une subversion*. Paris: Syros.

Van der Loo, Hans, Erik Snel, and Bart van Steenbergen. 1984. *Een wenkend perspectief*. Amersfoort: De Horstink.

Veen, Hans-Joachim, and Jürgen Hoffmann. 1992. *Die Grünen: Zu Beginn der neunziger Jahre*. Bonn: Bouvier.

Vogel, David. 1988. Environmental policy in Europe and Japan. In Norman Vig and Michael Kraft, eds., *Environmental policy in the 1990s: Toward a new agenda*. Washington, D.C.: CQ Press.

————. 1993. Environmental policy in Europe. In Sheldon Kamieniecki, ed., *Environmental politics in the international arena*. Albany: SUNY Press.

Walker, Jack. 1991. *Mobilizing interest groups in America*. Ann Arbor: University of Michigan Press.

Walsh, Edward. 1981. Resource mobilization and citizen protest in communities around Three Mile Island. *Social Problems* 26: 1–21.

————. 1989. *Democracy in the shadows*. Boulder, Colo.: Greenwood Press.

Walsh, Edward, and Rex Warland. 1983. Social movement involvement in the wake of a nuclear accident. *American Sociological Review* 48: 764–80.

Warren, Karen. 1987. Feminism and ecology: Making connections. *Environmental Ethics* 9: 3–20.

Weber, Max. 1978. *Economy and society,* 2 vols. Berkeley: University of California Press.

Weiner, Martin. 1981. *English culture and the decline of the industrial spirit, 1850–1980*. Cambridge: Cambridge University Press.

Wey, Klaus. 1982. *Umweltpolitik in Deutschland: Kurze Geschichte des Umweltschutzes in Deutschland seit 1900*. Opladen: Westdeutscher.

Whiteman, David. 1990. The progress and potential of the Green Party in Ireland. *Irish Political Studies* 5: 45–58.

Wiesenthal, Helmut. 1993. *Realism in green politics: Social movements and ecological reform in Germany*. Manchester, Eng.: Manchester University Press.

Wilson, Frank. 1987. *Interest group politics in France*. New York: Cambridge University Press.

———. 1990. Neo-corporatism and the rise of new movements. In Russell Dalton and Manfred Kuechler, eds., *Challenging the political order*. New York: Oxford University Press, Cambridge, Eng.: Polity Press.

Witherspoon, Sharon, and Jean Martin. 1992. What do we mean by green? In Roger Jowell, ed., *British social attitudes: The ninth report*. Aldershot, Eng.: Dartmouth.

Young, K. 1991. Shades of green. In Roger Jowell, ed., *British social attitudes: The eighth report*. Aldershot, Eng.: Dartmouth.

Zald, Mayer, and Roberta Ash. 1966. Social movement organizations. *Social Forces* 44: 327–41.

Zald, Mayer, and John McCarthy, eds. 1987. *Social movements in an organizational society*. New Brunswick, N.J.: Transaction.

Index

Danish Association for Nature Conservation (DN), 31, 41, 61, 79, 86, 89, 90, 91, 93, 115, 190, 274n

Danish Ornithological Association (DOF), 30, 169, 274n

Democratic attitudes, 129–32, 207, 209

Diagnostic framing, 13, 130

Diani, Mario, 90, 220

Donati, Paolo, 99

Downey, Gary, 108

Downs, Anthony, 52

Dutch Bird Protection Association, 29

Earth First!, 11, 152

Ecology groups, 38–41, 119–21, 128, 131–32, 133, 141–42, 148, 157, 171, 174, 204, 223, 231, 233–34, 239, 246, 250–55, 258–59; definition of, 47–48, 78; organizational structure, 107, 108, 109; political actions, 144–45, 179, 195, 206–9, 223–24, 231, 252; political organizations, 16–17; public attitudes toward, 59–64; resources, 86, 96–97, 99

Education, 67, 71, 116

Electoral Politics, 232–40

Environmental disasters, 36, 43–44

Environmental movement: history of, 26–38; present, 38–41, 244–49, 256–59

Environmental orientations, 46–47, 113, 272n. See also Conservation groups; Ecology groups

EREYA, 82, 256

Eurobarometer survey, 18, 52, 53, 56, 57, 60, 159, 235

Eurocommunist parties, 224, 234

European Community, 18, 248, 249

European Community Parliament, 44, 116

European Environmental Bureau (EEB), 172, 249, 261–62

Fauna and Flora Preservation Society (FFPS), 114, 271n

Federal Association of Citizen Action Groups for Environmental Protection (BBU), 41, 81, 220, 248

Forschungsgruppe Wahlen, 54

Foss, Daniel, 10

Foundation for Alternative Living, 40, 79, 92, 245

Foundation for Environmental Education (SME), 40, 81, 169

Foundation for Nature and the Environment (SNM), 81, 172, 190, 195

Frankland, Gene, 90

French Federation of Societies for the Protection of Nature (FFSPN), 31, 41, 81, 89, 91, 170, 172

Friends of the Earth (FoE), 39–40, 90–91, 107, 109, 169, 220, 245, 255; Belgium, 39, 219; Britain, 39, 90, 93, 137, 194, 196, 211, 219, 271n, 281n, 283n; France, 39, 219, 283n; Greece, 39; Italy, 39, 42, 219; Ireland, 39; Netherlands, 39, 91, 216; United States, 39, 169

Friluftradet, 81, 171, 191, 280n

Fuchs, Dieter, 66

Fund for the Italian Environment (FAI), 41

Gallup Health of the Planet Study, 59, 273n

Gelb, Joyce, 200

Gerlach, Luther, 110, 246

German Federation for Bird Protection (DBV), 29, 34, 41, 86, 192, 250

German Federation for Environmental and Nature Protection (BUND), 192

German Green party, 9, 48, 115, 161, 236, 283n

German Nature Protection Ring (DNR), 31, 34, 81, 89, 172, 280n

Goyder, Jane, 47, 84, 88, 141, 173, 220, 262

Green Alliance, 82, 167, 171

Green/New Left parties, 59, 102, 154, 156, 157, 174, 212, 215, 218–19, 221–22, 224, 228, 230, 233–34, 255

Greenpeace, 40, 42–43, 82, 87, 90–91, 107–8, 115, 139, 146, 152, 169, 196, 232, 236, 245, 250, 254, 255; Belgium, 40; Britain, 40, 93, 107, 108, 111, 115, 181, 188; Denmark, 40; France, 40, 81, 91; Germany, 40, 93; Ireland, 40; Italy, 40, 115; Luxembourg, 40; Netherlands, 40, 86, 91, 93, 115, 170, 186, 189

Old Left, 126, 214, 237
Old Politics, 126–29
Olson, Mancur, 89
Opportunity structures, 14, 200, 205–6, 214, 230, 231
Organizational structure, 7, 9, 99–108, 109, 199, 204–5, 252–53

Papandreou, Georges, 161, 228
Parliaments, 156, 160–61, 174, 206, 280*n*
Peace movement, 102, 128, 168, 200, 212, 240, 279*n*
Personnel (staffing) of groups, 97–99
Piven, Frances, 110
Plumage League, 29
Plumage trade, 28–30
Policy activity, 143–48, 178–79
Political activities, 173–76, 178–85, 198–209, 227–32, 245–46, 252–54, 258. *See also* Conventional political action; Protest activities
Political parties, 213–40. *See also specific party groups*
Pompidou, Georges, 38
Porritt, Jonathon, 211, 218, 277*n*
Postmaterialism, 4, 64–66, 67, 127, 273*n*
Prognostic framing, 13, 200
Pro Natura, 42
Protest activities, 184, 195–98, 202–9, 252, 257
Public opinion, 53, 55, 184; salience of issue, 51–52, 53–54; support for environment, 4, 55–59, 216, 243; support for groups, 59–64

Raad Leefmilieu te Brussel, 81, 192
Rational Choice Theory. *See* Resource Mobilization Theory
Reif, Karlheinz, 55
Resource Mobilization Theory (RM), 5–8, 9, 10, 11, 15, 84–85, 110, 150, 151, 173, 178, 200, 230–31, 246, 249–50, 251, 254–55, 269*n*, 276*n*
Resources, 92–97, 199, 202–4, 229–30, 247; sources, 94–97, 203, 275*n*
Ridley, Nicholas, 121
Rio Summit, 45, 244

Robinwood, 11, 13, 41, 92, 97, 169, 196, 248
Rochon, Thomas, 212
Rohrschneider, Robert, 65, 66
Royal Society for Nature Conservation (RSNC), 86
Royal Society for the Prevention of Cruelty to Animals (RSPCA), 28, 169
Royal Society for the Protection of Birds (RSPB), 14, 29, 41, 86, 89, 90, 93, 97, 114, 152, 169, 188, 204, 245, 251
Rucht, Dieter, 179, 198, 201, 230

Salisbury, Robert, 100
Schumacher, E. F., 47, 110
Schutzgemeinschaft Deutscher Wald, 192
Selbourne Society, 29, 169
Sellafield nuclear power facility, 137, 180–81
Sizewell, commission on nuclear power, 194
Snow, David, 13, 138, 200, 251
Social class, 65, 69, 115, 116–17
Social Democratic Parties, 53, 216, 225
Socialist Environment and Resource Association, 166, 217, 283*n*
Socialist parties, 156, 216, 222, 224, 226, 228
Social Movement Organization (SMO), 6, 7, 9, 111, 150, 187, 226, 248, 249–50, 251, 254
Society for the Defense of the Environment (VMD). *See* Friends of the Earth, Netherlands
Society for the Promotion of Nature Reserves (SPNR), 30, 170
Society for the Protection of Nature (SNPN), 31, 83, 115
SPIN networks, 247
Stichting Leefmilieu, 81, 172
Student protests, 36–37
Swiss League for the Protection of Nature, 34

Tarrow, Sidney, 215
Thatcher, Margaret, 42, 44, 161–62, 217
Tilly, Charles, 10, 150, 168, 254
Times of London, 28, 137, 180, 188

Town and Country Planning Association
(TCPA), 33
Tucker, William, 114
Turner, Ralph, 179

Walker, Jack, 14
Walsh, Edward, 201
Weber, Max, 100
Wildlife Link, 171
Windscale nuclear power facility. *See* Sellafield nuclear power facility

Women's movement, 102, 115–16, 127, 168, 200, 240, 269*n*, 279*n*
World Wildlife Fund (WWF), 41–42, 90–91, 95, 115, 245; Belgium, 35; Britain, 35, 93, 115; Denmark, 35, 169; France, 35; Germany, 35, 89, 93; Italy, 35, 90, 93, 115, 220; Netherlands, 35, 91, 115, 169

Zimmermann, Friedrich, 159